Index
to
Sections

...mercial Fluorescent, Industrial Fluorescent, Commercial-...HID, Commercial Incandescent, Industrial Incandescent, ...l-Industrial Emergency, Residential Incandescent
Exterior: Residential, Commercial Building, Floodlighting and Streetlighting

CIRCUIT PROTECTION
Bolted Pressure Switches, Distribution Panelboards, Branch Circuit Panelboards, Safety Switches, Circuit Breakers, Molded-Case Switches

SERVICES
Service Entrance, Current Transformer Cabinets, Multimetering, Service Gutter or Wiring

GROUNDING
Ground Rods, Ground Clamps, Grounding Conductors, Exothermic Connections, Lightning Protection

FEEDERS
Conduits, Conduit Fittings, Conductors, Conductor Terminals and Taps, Feeder Busway, Plug-in Busway, Duct Banks, Ladder Trays

BRANCH CIRCUITS
Conduit—with Wire, Conduits—Empty, Conduit Fittings, Conductors, Mineral-Insulated Cable, Outlets, Appliance Connections, Motor Terminal Connections, Surface Raceways, Undercarpet Wiring System, Underfloor Raceway System, Trench Duct System, Lighting Duct System

CONTROL EQUIPMENT
Low-Voltage Control System, Photoelectric and Time Switches, Multipole Relays, Lighting Contactors, Dimmers, Motor Starters, Motor Control Centers

TRANSFORMERS
Dry Type, Oil Type

AUXILIARY SYSTEMS
Telephone, Signaling, Clock/Program, Fire Alarm, Nurse Call, Sound, Emergency Call, Apartment Intercom, Master Television Antenna

ELECTRIC HEATING AND CONTROLS
Ceiling or Suspended Type, Wall Type, Freeze Protection, Floor Heating, Snow Melting, Controls

POWER DISTRIBUTION ABOVE 600 VOLTS
Overhead, Underground

MISCELLANEOUS
Junction Boxes, Trenching and Backfilling, Drilling Holes, Channeling for Conduit, Cutting Pavement, Anchors

APPENDIX
Sample Pricing and Summary Sheets, Sample Problems

Rapid Electrical Estimating and Pricing

Rapid Electrical Estimating and Pricing

A HANDY, QUICK METHOD OF DIRECTLY DETERMINING THE SELLING PRICES OF ELECTRICAL CONSTRUCTION WORK

C. Kenneth Kolstad, P.E.
Gerald V. Kohnert, P.E.

Fourth Edition

McGraw-Hill Book Company   New York   St. Louis   San Francisco   Auckland   Bogotá   Hamburg

London   Madrid   Mexico   Montreal   New Delhi   Panama   Paris   São Paulo   Singapore

Sydney   Tokyo   Toronto

Library of Congress Cataloging in Publication Data

Kolstad, C. Kenneth.
   Rapid electrical estimating and pricing.

   Includes index.
   1. Electric engineering—Estimates.   I. Kohnert,
Gerald V.   II. Title.
TK435.K6 1986   621,319′24   85-11649

34567890   HAL/HAL   898

ISBN 0-07-035131-7

*The editors for this book were Harold B. Crawford and Beatrice
E. Eckes, the designer was Naomi Auerbach, and the production
supervisor was Sally Fliess. It was set in Press Roman Italic by
Florence Lanaro.*

Printed and bound by Halliday Lithograph.

Table of Contents

PREFACE   vii

HOW TO USE THIS BOOK   xi

HOW TO ADJUST FOR LABOR AND MATERIAL PRICE CHANGES   xiii

TOTAL PRICE ADJUSTMENT CHART   xiv

INDEX TO SECTIONS   xvii

SECTIONS A-L

APPENDIX

Preface

During our years in the electrical contracting industry and as consulting engineers, we realized that a method for establishing a cost estimate which was both reasonably accurate and quick was needed. Such a method of working out engineering estimates would be of considerable value to contractors, architects, engineers, physical-plant superintendents, and others.

Previously, a proper estimate would require more time than was economically feasible, but the only accurate method demanded an estimate similar to the one used by contractors. A takeoff of the items required with labor units assigned to the various pieces and, finally, a summary of the estimate would be necessary. Since this was time-consuming and costly and did not necessarily result in a sale, the time spent was often wasted. But this was about the only accurate method available.

For an estimate to be of value to the contractor or engineer, it should be possible to prepare it rapidly and with a reasonable degree of accuracy. A departure from the conventional material and labor-hour columns with the additions of overhead, labor burden, taxes, profit, etc., is necessary. All these items have to be in the "rapid" price. Thus, the qualified observer will notice that the graphs in this book show neither material prices nor labor hours so that a quick and accurate price in dollars can be determined.

The prices shown on the graphs include contractors' material costs, labor hours for installation at an overall national average of $16 per hour, overhead at an average of 85% on labor, direct job expenses of 12% on labor, and 5% profit on prime costs (material + labor + overhead + direct job expenses). Overhead includes administrative salaries, rent, bookkeeping, telephones, insurance, promotional expenses, etc., and costs not associated with a specific job, which generally vary inversely with the size of the firm. Direct job expenses as indicated here include time estimating, field shop, field telephone, job tools used and depreciated, job supervision, and freight and truck expenses relating to the specific job. As a consequence of these costs added to the material and labor, graph prices reflect the total installed selling price by the electrical contractor.

A Total Price Adjustment Chart which deals with the mechanics of adjusting labor rates and material prices is shown on page xiv. This is an extremely important aspect of estimating prices and must be understood by the user of this book before establishing any firm estimate.

This book will be of considerable value and assistance to many technical people involved in electrical construction work. Consulting engineers, electrical contractors, general contractors, architects, physical-plant superintendents, electrical equipment manufacturers, and electrical wholesale representatives serving smaller contractors will find the book useful, as will engineers in such government agencies involved in construction as the Army, Navy, Air Force, General Services Administration, and state and municipal engineering departments, and, finally, industrial, college, and university physical-plant administrators.

Since the first three editions of this book have been so widely accepted, many more items, allowing more thorough estimates, have been included in this fourth edition. Owing to technological progress, a number of items which are obsolete have been deleted. Many new items have been added and existing ones updated as seemed appropriate. The book should prove to be as satisfactory a working tool for the user as it has been for those of us who have been using this method over the years.

The book is organized into sections corresponding to the same general categories as an electrical construction project. These categories are: Lighting, Circuit Protection, Services, Grounding, Feeders, Branch Circuits, Control Equipment, Transformers, Auxiliary Systems, Electric Heating and Controls, Power Distribution above 600 Volts, and Miscellaneous. The Appendix contains four typical problems with their complete solution to demonstrate the manner of rapidly finding the job selling price.

It has been found that a job can be taken off and priced in about one-fifth of the time required by the conventional system with the same degree of accuracy as you find between bidders. It might be pointed out that it is rather difficult to evaluate bids correctly when there are only two bidders, particularly if one is not familiar with the bidding situation in the areas involved at the time of bidding. Five bidders provide a much clearer picture. When one is familiar with conditions in an area, the total price adjustment multiplier that is more accurate for one's use can be easily found. We have also found that on large jobs where there are large open spaces or much repetition of lighting luminaires and panelboards there often is competition between manufacturers and wholesalers which affects the accuracy of estimates from this book. The estimator must be aware of this situation and adjust to it as he or she sees fit.

We would like to express our appreciation to the following well-qualified people who, through their own specialized knowledge, have given help and suggestions for this project:

Allen L. Bader *President, Electrical Construction Co., Colorado Springs, Colo.*

Clarence (Clancy) O. Bader *Electrical Construction Co., Colorado Springs, Colo.*

Sandy Batten *Gregg Cloos Co., Denver, Colo.*

William J. Birkett *Manager, Southern Colorado Chapter, National Electrical Contractors Association, Colorado Springs, Colo.*

Richard G. Carroll *Graybar Electric Co., Inc., Colorado Springs, Colo.*

Dale Chase *Berwick Electric Co., Colorado Springs, Colo.*

James Cooper *American Electric Co., Colorado Springs, Colo.*

G. T. (Jerry) Daveline *Square D Company, Pueblo, Colo.*

David J. Evancheck *Simplex Time Recorder Co., Denver, Colo.*

Douglas J. Franklin *Thompson Lightning Protection, Inc., St. Paul, Minn.*

Steve Goldstein *Cadweld Representative, Denver, Colo.*

Bob Hughes *Electro-Media of Colorado, Denver, Colo.*

David Marquardt *Manufacturers' Representative, Denver, Colo.*

Allan F. Nies *Manufacturers' Representative, Denver, Colo.*

Larry J. Peifer *Electrical Construction Co., Colorado Springs, Colo.*

Richard B. Schumann *The Wiremold Co., Denver, Colo.*

Craig D. Smith *Thomas & Betts Co., Denver, Colo.*

L. Kay Waldron, P.E. *Owner, Adams Excavating Co., Colorado Springs, Colo.*

C. Kenneth Kolstad

and

Gerald V. Kohnert

How to Use This Book

The basic concept of this book of graphs is to provide a means of quickly pricing a job with a reasonable degree of accuracy. The use of assemblies, as developed by the Estimatic Corporation, was followed and modified where required. For example, a 60-ampere, 250-volt, three-phase, four-wire safety switch, NEMA 1, in place ready for use, consists of:

1. One 60-ampere switch
2. Three fuses (Fusetrons are used here)
3. Four wall fastenings (toggle bolts are used here on the assumption of mounting to a masonry wall)
4. Twelve feet of #8 THW wire (copper) in the switch and eight wire terminations
5. Two 1-inch galvanized rigid conduit terminals, which include cutting and threading the conduit and installing two locknuts and a metal-insulated bushing on each conduit terminal

Obviously a number of the items indicated could be changed, using the same switch, but the price change would be insignificant in relation to the whole assembly price, including labor, overhead, direct job expense, and profit; hence an average finite value can be used with reasonable accuracy.

In view of the above being typical of assemblies, it is important to become familiar with the description of each graph in order to understand all that each graph implies. Since these graphs are not intended to furnish a job, as an NECA or Estimatic takeoff, certain liberties can be taken in the interests of speed. If a contractor should sell a job from these graphs, he or she can then make a material takeoff to furnish the job, since the contractor can then afford the time to do this work.

The use of these graphs can be hazardous for the inexperienced estimator because they are based upon average conditions. Job factors must be considered and applied as required. For example, labor units are not the same when conduit must be installed for a 25-foot-high ceiling as for a 12-foot-high ceiling. Labor units for installing fixtures from a raw dirt floor will not be the same as from a clean concrete slab. The estimator must be cognizant of the many pitfalls that can turn an average job into a loser. Some geographical areas also seem to have a planned productivity control which must also be evaluated by the estimator.

Occasionally, it is required to relocate existing equipment such as safety switches, panelboards, feeder, or plug-in busway. Fixtures are sometimes supplied by the owner to the contractor. In these cases it is necessary to deduct only the cost of the equipment. Fastening devices, lamps, connectors, wires, hangers, etc., and their associated labor as well as overhead and direct job expense are still required. In order to separate out the cost of equipment items, it is only necessary to procure costs of materials from a wholesaler or a catalog, then deduct this amount for the equipment only. The only costs will remain in the estimate, as they should.

A departure from the normal pricing method has been introduced on pages A-21, A-22, and A-23 relating to residential fixtures. It has been our experience that many residential jobs are done with an allowance provided for the cost of fixtures. Obviously, the owner can select residential fixtures with a wide cost variation. However, the installation cost will not vary greatly in most instances. To allow the customer to provide whatever he or she wants, you, the contractor, are expected to install the fixture. In this respect we have selected a few typical residential fixtures, assigned labor units and lamps, then determined the selling price for the installation only, which recovers direct labor, overhead, direct job expense, and some profit.

The sections of the book have been arranged so that an orderly progression through a complex job can be made with ease and with some semblance of order. There are many ways to make a takeoff, all of which are satisfactory, and the following is just one suggestion. On pricing and summary sheets like those shown in the examples in the Appendix:

1. List quantity of each type of fixture.
2. List distribution equipment and panelboards.
3. List service equipment.
4. List feeder conduits, conductors, fittings, etc.
5. List branch circuit outlets, receptacles, switches, conduits, wire, etc.
6. List motor-feeder disconnects, starters, controllers, connections to motors and appliances, dry transformers, etc.
7. List auxiliary system equipment such as telephone panels, fire alarm equipment, audible signals, sound system, etc.
8. With equipment and other items listed on the pricing sheets, look up the item costs, extend, and summarize.
9. After summarizing the costs on the summary sheet, apply the price adjustment multiplier as determined from the Total Price Adjustment Chart. Apply the multiplier to come up with a total sell price shown at the bottom of the summary sheet. (See Appendix.)

Occasionally you are not able to locate in REEP an item that is required on the job. (Refer to Sheet 2 of Example 1 in the Appendix.) To handle this sort of problem, insert the item near the bottom of the page of the pricing sheet under "Description" where "phase loss relay" is listed. Associate the unit material costs and labor units as shown.

The reader is referred to examples of pricing in the Appendix. Example 1 illustrates pricing of a distribution system from a one-line diagram which was made up for illustrative purposes only. Example 2 illustrates the pricing of a main distribution panel and motor control center. Example 4 illustrates the pricing of a clock/sound system installed in an existing school.

Spare blank forms are provided at the end of the book so that you may cut them out and have your printer copy as many as you need.

How to Adjust for Labor and Material Price Changes

Probably the most generally asked question regarding the first two editions has been related to compensation for labor and material price changes of the marketplace. The method we have developed has worked quite well.

Since we are dealing with averages in the establishment of the unit costs, we must be able to adjust for specific geographical areas. To do this we make the assumption that the amount a worker is paid does not influence the worker's productivity—that the worker installs the same amount of material. If the quantity of work installed in a day is further controlled, the labor productivity factor should be adjusted up. The authors have assumed and corroborated that journeymen require about 30 minutes in the morning and afternoon of nonproductive time for coffee breaks, relief time, tool and material acquisition, and cleanup. Since this time totals an hour per day, it is necessary to add this cost, which amounts to a 12 1/2% adder, to determine the total hourly labor cost.

To use the Total Price Adjustment Chart properly, it is necessary to compensate for both labor and material.

TOTAL PRICE ADJUSTMENT CHART

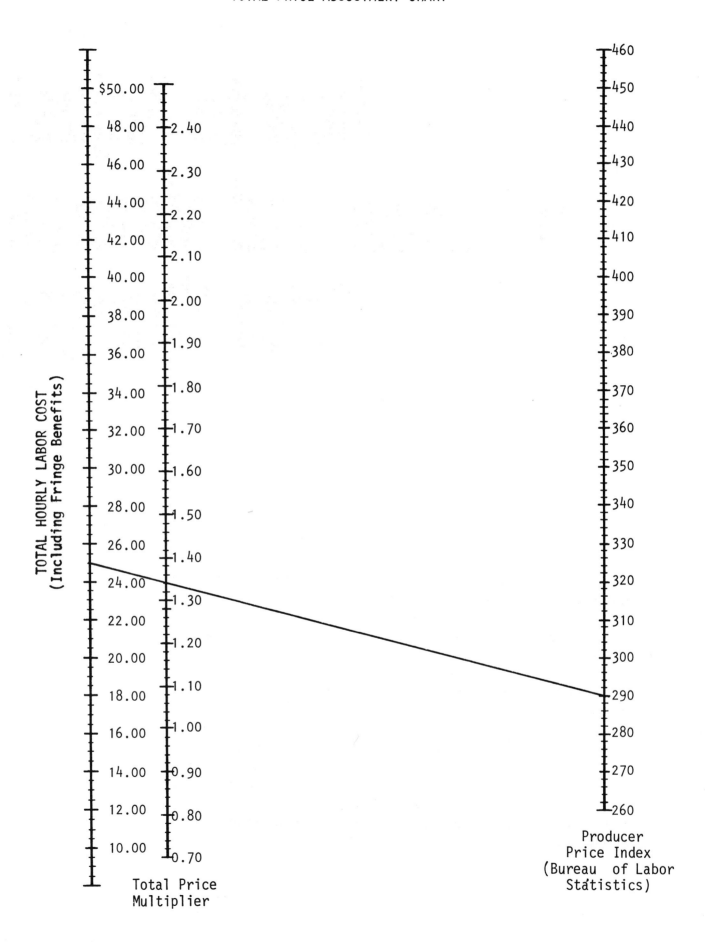

TOTAL HOURLY LABOR COST
(Including Fringe Benefits)

Total Price
Multiplier

Producer
Price Index
(Bureau of Labor
Statistics)

TOTAL HOURLY LABOR COST

	Typical		Commercial industrial labor rate, your area		Light commercial labor rate, your area		Residential labor rate, your area	
	Percent	Dollars	Percent	Dollars	Percent	Dollars	Percent	Dollars
Base rate per hour	—	$16.00						
FICA (social security)	7.05%	1.13						
State unemployment tax	5.40	.86						
Federal unemployment tax	6.20	.99						
State workmen's compensation insurance and general liability insurance	5.75	.92						
Health and welfare fund	—	1.25						
Pension Fund	—	1.00						
National Electrical Benefit Fund	3.00	.48						
National Electrical Contractors Association	0.80	.13						
Apprentice Fund JATC	—	.06						
National Electrical Industry Fund	1.00	.16						
Labor productivity factor (two coffee breaks, tool acquisition, reliefs,etc.)	12.50	2.00						
Total hourly labor cost		$24.98						

Labor adjustment: Fill in the appropriate figures in the above list to determine the total hourly labor cost to be applied to the specific job at hand. Bear in mind that these items change with area and time.

Material adjustment: Look in the magazine U.S. News and World Report under U.S. Business and the chart USN & WR Weekly Index of Business Activity, Monthly Indicators, Producer Prices, Finished Goods, and find the latest price index. Also available is the Monthly Labor Review, published by the Bureau of Labor Statistics, which is available in most libraries.

Total price adjustment: Draw a line from the total hourly labor cost to the latest producer price index and note the total price multiplier. Use this to multiply by your total takeoff cost from the graphs. Let's say that you find the producer price index is 290.3; use 290 and go to the Total Price Adjustment Chart on the right side. Since the total hourly labor cost is $24.98, use $25.00, draw a line from $25.00 to 290, and find the total price multiplier of 1.34. After you have summarized your complete takeoff, multiply the total by 1.34, which gives the total sell price.

Index to Sections

LIGHTING

Interior: Commercial Fluorescent, Industrial Fluorescent, Commercial-Industrial HID, Commercial Incandescent, Industrial Incandescent, Commercial-Industrial Emergency, Residential Incandescent
Exterior: Residential, Commercial Building, Floodlighting and Streetlighting

CIRCUIT PROTECTION

Bolted Pressure Switches, Distribution Panelboards, Branch Circuit Panelboards, Safety Switches, Circuit Breakers, Molded-Case Switches

SERVICES

Service Entrance, Current Transformer Cabinets, Multimetering, Service Gutter or Wiring

GROUNDING

Ground Rods, Ground Clamps, Grounding Conductors, Exothermic Connections, Lightning Protection

FEEDERS

Conduits, Conduit Fittings, Conductors, Conductor Terminals and Taps, Feeder Busway, Plug-in Busway, Duct Banks, Ladder Trays

BRANCH CIRCUITS

Conduit–with Wire, Conduits–Empty, Conduit Fittings, Conductors, Mineral-Insulated Cable, Outlets, Appliance Connections, Motor Terminal Connections, Surface Raceways, Undercarpet Wiring System, Underfloor Raceway System, Trench Duct System, Lighting Duct System

CONTROL EQUIPMENT

Low-Voltage Control System, Photoelectric and Time Switches, Multipole Relays, Lighting Contactors, Dimmers, Motor Starters, Motor Control Centers

TRANSFORMERS

Dry Type, Oil Type

AUXILIARY SYSTEMS

Telephone, Signaling, Clock/Program, Fire Alarm, Nurse Call, Sound, Emergency Call, Apartment Intercom, Master Television Antenna

ELECTRIC HEATING AND CONTROLS

Ceiling or Suspended Type, Wall Type, Freeze Protection, Floor Heating, Snow Melting, Controls

POWER DISTRIBUTION ABOVE 600 VOLTS

Overhead, Underground

MISCELLANEOUS

Junction Boxes, Trenching and Backfilling, Drilling Holes, Channeling for Conduit, Cutting Pavement, Anchors

APPENDIX

Sample Pricing and Summary Sheets, Sample Problems

Lighting

INTERIOR/COMMERCIAL FLUORESCENT

Recessed—Lay-in Grid   A-1
Recessed—Flanged Type   A-2
Air Handling—Recessed—Lay-in Grid   A-3
Air Handling—Recessed—Flanged Type   A-4
Parabolic Loūver—Recessed—Grid & Flanged Type—Air Handling   A-5
Ceiling—Surface Mount—Metal-Sided   A-6
Ceiling—Surface Mount—Specification-Grade Wraparound   A-7
Ceiling—Surface for Damp and Wet Locations   A-7
Wall Mount—Over Mirror   A-7
Competitive Grade—Recessed, Wraparound, Surface   A-8
Fluorescent Strips   A-9

INTERIOR/INDUSTRIAL FLUORESCENT

Specification- and Competitive—Grade Industrials   A-10

INTERIOR/COMMERCIAL—INDUSTRIAL HID

Industrial—Specification-Grade & Decorative Shades   A-11
Recessed Rounds and Squares   A-12
Recessed—Lay-in Grid and Accent Units   A-13

INTERIOR/COMMERCIAL INCANDESCENT

Recessed Incandescent—Square   A-14
Recessed Incandescent—Rounds   A-14
Wall- and Ceiling-Mounted   A-15
Exit Lights   A-16
Lite-Trac   A-16

INTERIOR/INDUSTRIAL INCANDESCENT

Ceiling- and Wall-Mounted Globes and Reflectors   A-17

INTERIOR/COMMERCIAL—INDUSTRIAL EMERGENCY

Battery-Powered Emergency Lighting—Unit Type   A-18
Battery-Powered AC Systems   A-19
Battery-Powered Exit Light and AC System Accessories   A-20

INTERIOR/EXTERIOR RESIDENTIAL

Residential Incandescent Fixtures   A-21
Residential Incandescent Fixtures   A-22
Residential Incandescent Fixtures   A-23

EXTERIOR/COMMERCIAL BUILDING LIGHTING

Wall Brackets and Spots   A-24

POLES/FLOODLIGHTING AND STREET LIGHTING

Round Tapered Street Poles   A-25
Round Tapered Aluminum Poles   A-26
Square Tapered Steel Poles   A-27
Hinged Steel Poles   A-28
Light Duty—Round Tapered Aluminum and Steel Poles   A-29
Round Tapered Galvanized-Steel & Aluminum Poles
  with Pipe Arm   A-30
Round Tapered Galvanized-Steel & Aluminum Poles
  with Truss Arm   A-31
Poletop Brackets   A-32

EXTERIOR FLOODLIGHTING LUMINAIRES

Tungsten-Halogen   A-33
Mercury Vapor, Metal Halide, High-Pressure Sodium   A-33

EXTERIOR STREET AND AREA LIGHTING LUMINAIRES

Round and Square Post Top   A-34, A-35
Cutoff Luminaires   A-34, A-35
Fluorescent Area Light   A-34
Fluorescent Facade Light   A-34
Street-Lighting Luminaires, Post Tops   A-35
Bollard Lights   A-36

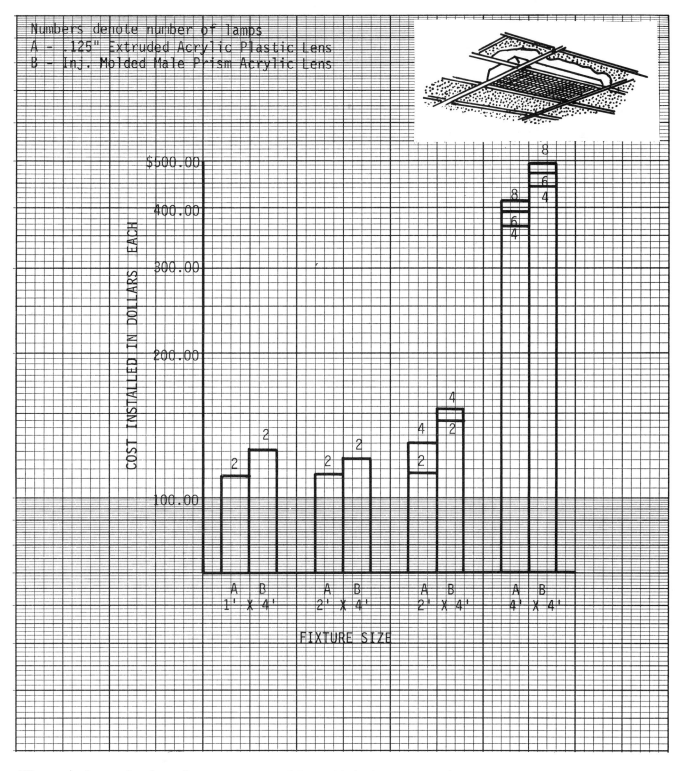

The costs shown for these fixtures consist of the published contractors' book price for the fixtures with the lenses shown, the number of lamps shown, and the labor required for the installation and connections of the fixture, including flexible cable. If an outlet for the fixture is required, see the Branch Circuit Section for ceiling outlet recessed above the ceiling. Reference group—Daybrite Designer.

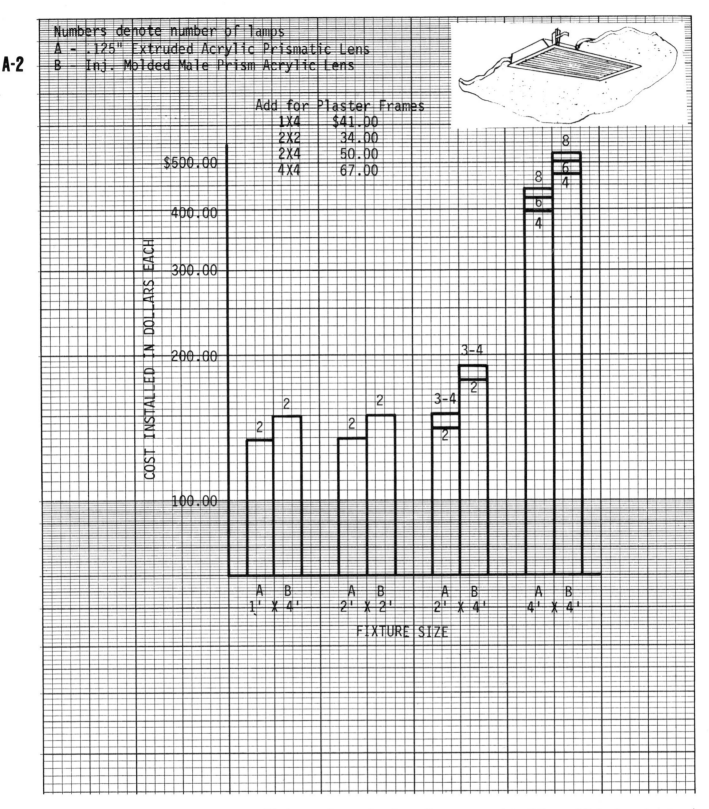

Numbers denote number of lamps
A - .125" Extruded Acrylic Prismatic Lens
B - Inj. Molded Male Prism Acrylic Lens

A-2

Add for Plaster Frames
1X4	$41.00
2X2	34.00
2X4	50.00
4X4	67.00

COST INSTALLED IN DOLLARS EACH

$500.00

400.00

300.00

200.00

100.00

A B   A B   A B   A B
1' X 4'   2' X 2'   2' X 4'   4' X 4'

FIXTURE SIZE

The costs shown for these fixtures consist of the published contractors'
book price for the fixtures with the lenses shown, the number of lamps
shown, and the labor required for the installation and connection of
the fixture, including flexible cable. If an outlet for the fixture is re-
quired, see the Branch Circuit Section for ceiling outlet recessed above
the ceiling. Reference group—Daybrite Designer.

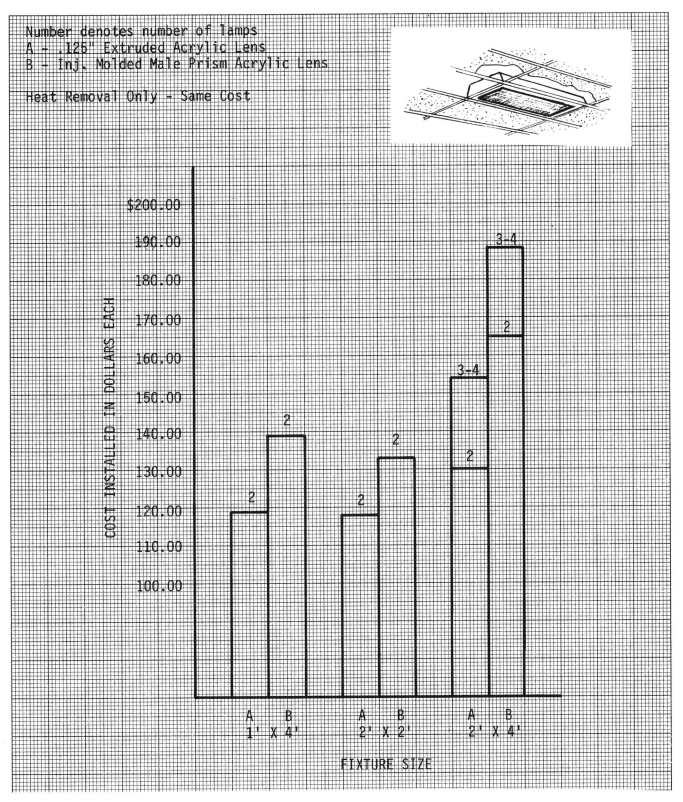

Number denotes number of lamps
A - .125" Extruded Acrylic Lens
B - Inj. Molded Male Prism Acrylic Lens

Heat Removal Only - Same Cost

COST INSTALLED IN DOLLARS EACH

$200.00
190.00
180.00
170.00
160.00
150.00
140.00
130.00
120.00
110.00
100.00

A   B
1' X 4'

A   B
2' X 2'

A   B
2' X 4'

FIXTURE SIZE

*The costs shown for these fixtures consist of the published contractors'
book price for the fixtures with the lenses shown, the number of lamps
shown, and the labor required for the installation and connections of
the fixture, including flexible cable. If an outlet for the fixture is required, see the Branch Circuit Section for ceiling outlet recessed above
the ceiling. Reference group—Daybrite Designer.*

A-4

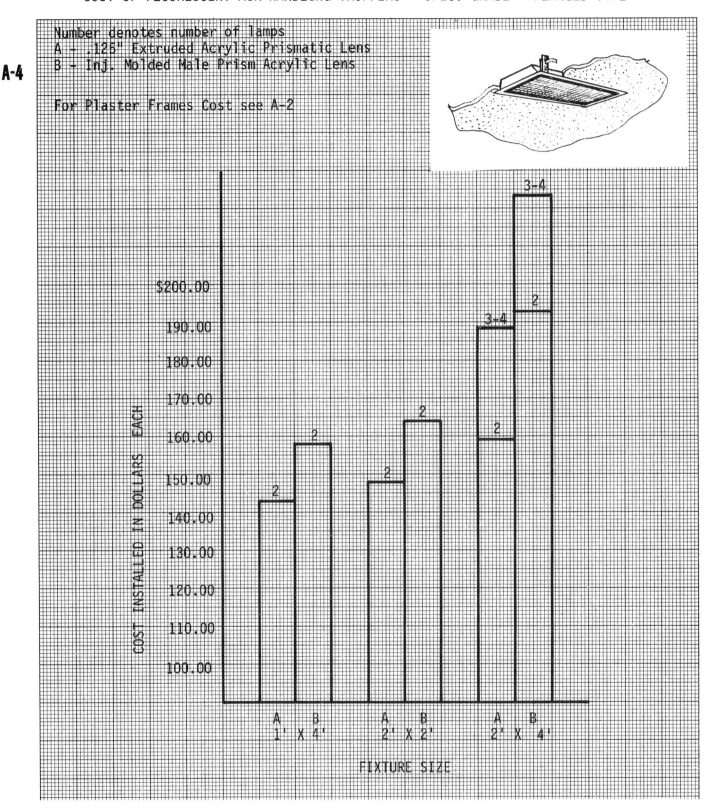

Number denotes number of lamps
A - .125" Extruded Acrylic Prismatic Lens
B - Inj. Molded Male Prism Acrylic Lens

For Plaster Frames Cost see A-2

COST INSTALLED IN DOLLARS EACH

$200.00
190.00
180.00
170.00
160.00
150.00
140.00
130.00
120.00
110.00
100.00

A	B	A	B	A	B
1' X 4'		2' X 2'		2' X 4'	

FIXTURE SIZE

*The costs shown for these fixtures consist of the published contractors'
book price for the fixtures with the lenses shown, the number of lamps
shown, and the labor required for the installation and connection of
the fixture, including flexible cable. If an outlet for the fixture is required, see the Branch Circuit Section for ceiling outlet recessed above
the ceiling. Reference group—Daybrite Designer.*

COST OF PARABOLIC LOUVER FLUORESCENT AIR-HANDLING TROFFERS

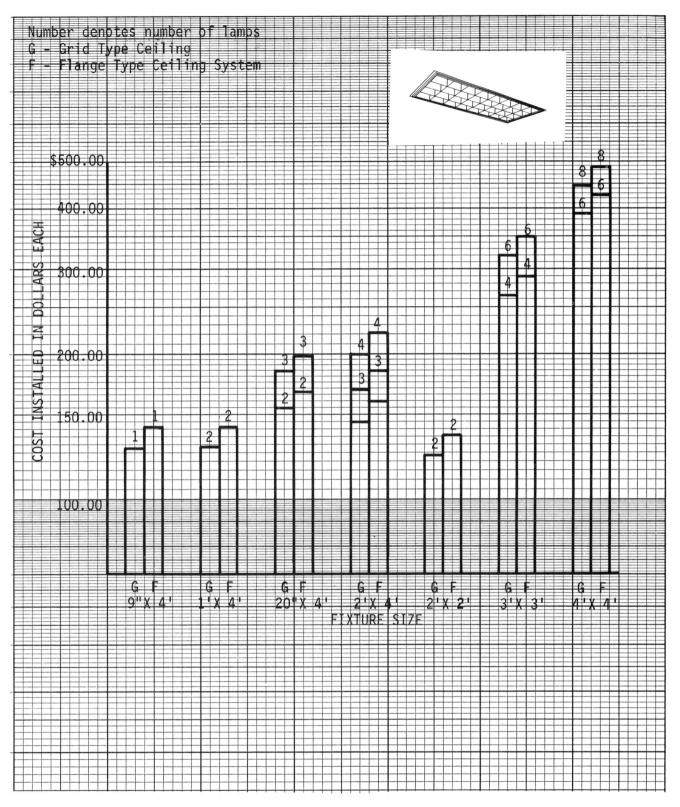

Number denotes number of lamps
G - Grid Type Ceiling
F - Flange Type Ceiling System

COST INSTALLED IN DOLLARS EACH

$500.00
400.00
300.00
200.00
150.00
100.00

| G F | G F | G F | G F | G F | G F | G F |
| 9"X 4' | 1'X 4' | 20"X 4' | 2'X 4' | 2'X 2' | 3'X 3' | 4'X 4' |

FIXTURE SIZE

*The costs shown for these fixtures consist of the published contractors'
book price for the fixtures with the lenses shown, the number of lamps
shown, and the labor required for the installation and connection of
the fixture, including flexible cable. If an outlet for the fixture is required, see the Branch Circuit Section for ceiling outlet recessed above
the ceiling. Reference group—Columbia Parabolume P-3 Series.*

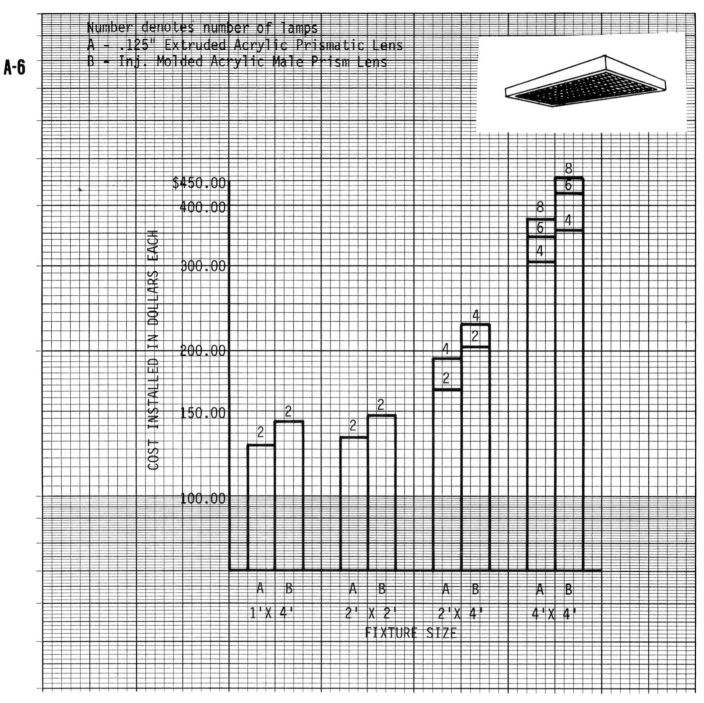

A-6

Number denotes number of lamps
A - .125" Extruded Acrylic Prismatic Lens
B - Inj. Molded Acrylic Male Prism Lens

The costs for these fixtures consist of the published contractors' book price for the fixtures with the lenses shown, the number of lamps shown, the required fixture fastening devices, and the labor required for the installation and connection of the fixture.

If an outlet is required, see the Branch Circuit Section for the proper type of outlet.

The 2' X 2' fixture is considered to use U lamps instead of the 20-watt trigger start lamps. Reference group—Daybrite Daylume.

Type of Unit	Manufacturer's Catalog No.	Description	Installed Cost
Daybrite - Hallmark	Surface Mount	Rapid Start	
	45257-4	2 lt.-4 ft.	$118.00
	45257-8	4 lt.-8 ft.	225.00
	45457-4	4 lt.-4 ft.	177.00
	Pendant Mount	(Stem not incl.)	
	45267-4	2 lt.-4 ft.	138.00
	46267-8	4 lt.-8 ft.	251.00
	45467-4	4 lt.-4 ft.	202.00
Vandal-Resistant	Surface Mount	Rapid Start	
	TV11240-4	2 lt.-4 ft.	101.00
	TV16440-4	4 lt.-4 ft.	148.00
For 24" single-stem hangers - Add:			13.00 each
Daybrite - Vaporlume	Surface Mount		
	Damp-R-41241	2 lt.-4 ft. R.S.	123.00
	Wet-WR-41241	2 lt.-4 ft. R.S.	135.00
	Damp-R-91241	2 lt.-8 ft. S.L.	268.00
	Wet-WR-91141	2 lt.-8 ft. S.L.	288.00
	Damp-R-81241	2 lt.-8 ft. H.O.	290.00
	Wet-WR-81241	2 lt.-8 ft. H.O.	310.00
Utility Units - Daybrite	Wall Mount		
	A 5220W BWE	2 lt.-2 ft. T.S.	109.00
	A 5230W BWE	2 lt.-3 ft. R.S.	126.00
	A 5240W BWE	2 lt.-4 ft. R.S.	122.00
	A 5220S S.S.	2 lt.-2 ft. T.S.	149.00
	A 5230S S.S.	2 lt.-3 ft. R.S.	185.00
	A 5240S S.S.	2 lt.-4 ft. R.S.	199.00

*The costs shown for these fixtures consist of the published contractors'
book price for the fixtures with the lenses shown, the required fixture
fastening devices, and the labor required for installation and connection
of the fixture. All fixtures are priced with virgin acrylic lenses.*

*To provide an outlet for the fixture, see the Branch Circuit Section for
the type of wall or ceiling outlet required.*

A-8

Type of Unit	Manufacturer	Description	Installed Cost
Recessed in Grid .110" Prismatic Acrylic Lens	Lithonia 2GS2U40A12	Static Unit 2'X 2' - 2lt.	$ 99.00
	2GS240A12	2'X 4' R.S. 2 lt.	103.00
	2GS340A12	2'X 4' R.S. 3 lt.	114.00
	2GS440A12	2'X 4' R.S. 4 lt.	118.00
		For Heat Removal	Add $4.00 to each
Wraparound-Acrylic Lens		Rapid-Start Lamps	
	WA240A	2 lt. - 4 ft.	91.00
	8TWA440A	4 lt. - 8 ft.	176.00
	WA440A	4 lt. - 4 ft.	140.00
Surface Modular .110" Prismatic Acrylic Lens	2M2U40-MA12	2'X 2' - 2 lt.	119.00
	M240-MA12	1'X 4' - 2 lt.	110.00
	2M240-MA12	2'X 4' - 2 lt.	139.00
	2M440-MA12	2'X 4' - 4 lt.	156.00

The costs shown for these fixtures consist of the published contractors' book price for the fixture with the lenses shown, the number of lamps as shown, the required fixture fastening devices, and the labor required for the installation and connection of the fixture.

To provide an outlet for the fixture required see the Branch Circuit Section.

The illustrations shown are only to give a general idea of the type of fixture represented on the chart.

These fixtures are all of the competitive grade.

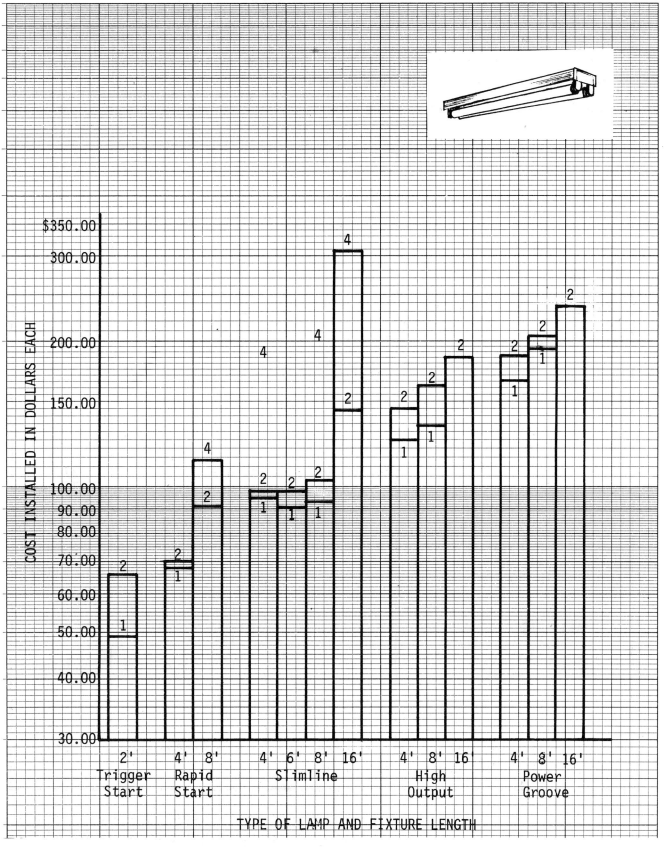

The costs shown for these fixtures consists of the published contractors'
book price for the fixture with the number of lamps as shown, the
required fastening devices, and the labor required for the installation
and connection of the fixture. To provide an outlet for the fixture, see
the Branch Circuit Section for the proper outlet.

A-10

Type of Unit	Manufacturer's Catalog Number	Description	Installed Cost
Specification Grade Daybrite CFI-25 Polyester Enamel	R-41252W-4	2 lt. - 4 ft. R.S.	$102.00
	R-41452W-8	4 lt. - 8 ft. R.S.	167.00
	R91252W-8	2 lt. - 8 ft. S.L.	150.00
	R-81252W-4	2 lt. - 4 ft. H.O.	129.00
	R-81252W-8	2 lt. - 8 ft. H.O.	172.00
	R-71252W-4	2 lt. - 4 ft. P.G.	177.00
	R-71252W-8	2 lt. - 8 ft. P.G.	219.00
Hinged louvers - with 30°C and 20°L shielding in white enamel	7954-4 Rapid-Start Slimline and High-Output Lamps	4' unit	35.00
		8' unit	71.00
	7955-4 For Power-Groove Lamps	4' unit	37.00
		8' unit	74.00
Competitive Grade Daybrite Dayline Baked white Enamel		2 lt. - 4 ft. R.S.	73.00
		2 lt. - 8 ft. S.L.	118.00
		2 lt. - 8 ft. H.O.	134.00
		2 lt. - 8 ft. P.G.	179.00
For 120-volt Hydee Hanger - Add $30.00			

The costs shown for these fixtures consist of the published contractors' book price for the fixture with the number of lamps indicated. No hanging or fastening materials are included for the suspension system, which must be priced separately.

Type of Unit	Catalog Number	Wattage & Lamp Type		Installed Cost
Bantam Unit - Open Reflector	BANT100MV	100W	M.V.	$201.00
	BANT175MV	175W	M.V.	218.00
	BANT250MV	250W	M.V.	231.00
	BANT175MH	175W	M.H.	261.00
	BANT070HP	70W	H.P.S.	274.00
	BANT100HP	100W	H.P.S.	279.00
	BANT150HP	150W	H.P.S.	284.00
Enclosed & Gasketed Petrolux Refractor for Wet/Hazardous Locations (Class 1 - Div.2)	1971-PD-120-CL	100W	M.V.	224.00
	1973-PD-120-CL	175W	M.V.	226.00
	1972-PD-120-CL	250W	M.V.	243.00
	1975-PD-120-CL	175W	M.H.	257.00
	1977-PD-120-CL	70W	H.P.S.	283.00
	1970-PD-120-CL	100W	H.P.S.	290.00
	1979-PD-120 CL	150W	H.P.S.	294.00
Prismpack II	PP2K400MV	400W	M.V.	248.00
	2PP2K400MV	400W twin	M.V.	450.00
	PP2K400MH	400W	M.H.	290.00
	2PP2K400MH	400W twin	M.H.	567.00
	PP2KC1000MH	1000W	M.H.	401.00
	PP2K250HP	250W	H.P.S.	325.00
	PP2K400HP	400W	H.P.S.	340.00
	EM	Emer. Qtz. Lite		112.00
	0300	Wire Guard		18.00
Decorative Shades	CYS-5	Small Round		86.00
	CYS-3	Large Round		100.00
	SQR-5	Small Square		105.00
	SQR-3	Large Square		105.00
	HEX-5	Small Hexagonal		102.00
	HEX-3	Large Hexagonal		102.00

The costs shown for these fixtures consist of published contractors' book price. These fixtures are all as manufactured by Holophane, and include ballast for pendant mounting. Outlet boxes or power hooks not included. Labor includes assembly, hanging, and connection. Lamps of the proper type and size are included.

A-12

Type of Unit	Catalog Number	Wattage & Lamp Type		Installed Cost
Open with Clear Aluminum Reflector	1224H7-75MV-362	75W	M.V.	$258.00
	1226H5- 100MV-662	100W	M.V.	261.00
	1238H9-175MV-850	175W	M.V.	303.00
	1238H10-250MV-862	250W	M.V.	335.00
	1226S6-70HPS-652	70W	H.P.S.	391.00
	1226S5-100HPS-652	100W	H.P.S.	414.00
	1226S8-150HPS-652	150W	H.P.S.	443.00
	1238S10-250HPS-835	250W	H.P.S.	579.00
Ellipsoidal Reflector Low-Brightness Cone	1057H5-100MV-734	100W	M.V.	304.00
	1060H9-175MV-744	175W	M.V.	337.00
	1060H10-250MV-744	250W	M.V.	367.00
	1062H11-400MV-845	400W	M.V.	438.00
	1061M9-175MH-835	175W	M.H.	408.00
	1062M11-400MH-846	400W	M.H.	473.00
	1057S6-70HPS-734	70W	H.P.S.	401.00
	1057S6-100HPS-734	100W	H.P.S.	423.00
	1057S8-150HPS-734	150W	H.P.S.	449.00
	1061S10-250HPS-835	250W	H.P.S.	571.00
Square Regressed Lens Horizontal Lamp-Full Refl.	1016H5100FE-M8	100W	M.V.	250.00
	1314H9175FE-M8	175W	M.V.	288.00
Round Regressed Lens Horizontal Lamp-Full Refl.	91H5100MV-M8	100W	M.V.	250.00
	1391H9175FE-M8	175W	M.V.	285.00
Round Regressed Lens Vertical Lamp	90H5100FE-M8	100W	M.V.	295.00
	93057175FE-M8	175W	M.V.	319.00
	93057H10250FE-M8	250W	M.V.	241.00

The costs for these fixtures consist of the published contractors' book price. These fixtures are manufactured by Prescolite and include potted ballast, prewired junction box, and appropriate lamp.

COST OF RECESSED HID FIXTURES

Type of Unit	Catalog Number	Wattage & Lamp Type	Installed Cost
Clear Conoid Reflector	Prescolite 1227H9-982	175W R40 M.V.	$309.00
Wall Wash Unit	Prescolite 1286H9-175MV-929	175W R40 M.V.	376.00
Adjustable Accent Unit	Prescolite 1275H9-175FE-760	175W R40 M.V.	359.00
2'X 2' Grid-Mounted Unit	Holophane Multi-Lume II		
	MULTI175MH	175W M.H.	208.00
	MULTI250MH	250W M.H.	229.00
	MULTI400MH	400W M.H.	241.00

The costs for these fixtures consist of the published contractors' book price. These fixtures include potted ballasts, prewired junction box, and appropriate lamp.

A-14

Type of Unit	Manufacturer	Wattage	Installed Cost
Recessed Rounds Black Milli-groove 	Prescolite 1204-900 1212-910 1220-920 1227-980	 50R20 75R30 150PAR38 300R40	 $ 88.00 92.00 104.00 161.00
Recessed Round "A" Lamp Downlight 	Prescolite 1222-262 1224-362 1225-462	 100W 150W 200W	 101.00 114.00 129.00
Recessed Square Relampalite Trim 	Prescolite 488HF-7 1015HF-7 1313HF-7 1313HM-7	 100W 150W 200W 300W	 71.00 89.00 99.00 100.00

The costs for these fixtures consist of the published contractors' book price. Included are prewired junction boxes, the lamps indicated, and labor for installation.

COST OF SURFACE COMMERCIAL INCANDESCENT FIXTURES

Type of Unit	Manufacturer	Wattage	Installed Cost
Ceiling Mount	Prescolite		
	1102-910	75R30	$ 75.00
	1105-920	150R40	82.00
	1108-930	300R40	150.00
	1122-910	75R40	89.00
	1125-920	150R40	94.00
	1128-930	300R40	164.00
	1170-900	2-50W	109.00
	1172-910	2-75R30	110.00
	1175-920	2-150R40	137.00
	1178-932	2-300R40	193.00
Wall Bracket	Prescolite		
	WB-2	100W	41.00
	WB-28	200W	65.00
Lite-Form	Prescolite		
	4040	100W	92.00
Acrylic Prismatic Drum	Holophane		
	CUBC060	2-60W	82.00
	CUBC075	2-75W	87.00

The costs for these fixtures consist of the published contractors' book price for the fixtures shwon. Included are the required lamps and labor for installation. No outlet is included. See the Branch Circuit Section for the appropriate type of outlet.

SURFACE COMMERCIAL INCANDESCENT FIXTURES

The costs shown consist of the published contractors' book price for the track and the fixtures shown including the labor for installation. Any single run of track must also be provided with an outlet box. See the Branch Circuit Section for the type required, and add to the cost shown. Each run must have a starter track section, as it provides the branch circuit connection capability which the joiner track sections do not have. Add the track cost to the total fixture cost to find the system cost.

The exit sign shown is a standard, non-self-powered unit and includes two lamps. For self-powered signs, see page A-20.

Type of Unit	Description	INSTALLED COST	
		Surface Mount	Recessed
Prescolite Two-Circuit	4' Starter Trac Section with Live End Feed-in (without fixture)	$ 64.00	$121.00
	8' Starter Trac Section and Live End Feed-in (without fixture)	99.00	182.00
	8' Joiner Trac Section and Coupler (without fixture)	98.00	177.00
	Basic Unit with 150PAR38 lamp (upper illus.)	72.00	----
	White Cylindrical Fixture with 300PAR38 lamp (middle illus.)	114.00	----
	Wall Washer with Q250DC qtz lamp (lower illus.)	115.00	----
Surface/Pendant Exit Sign Thinline - Profile Series with 50,000-hour lamps	Prescolite 75221 Top or End Mount	88.00	----
	Pendant-Mounted	112.00	----
	Replacement Lamp 20T6½ DC/IF	4.00	----

A-17

Type of Unit	Manufacturer's Catalog Number	Wattage	Installed Cost
Porcelain Lamp Receptacle			
Keyless	100	$ 12.25
Pull Chain	100	13.00
RLM Porcelain Reflector	Appleton		
	G50212	100	43.75
	G50214	150	46.25
	G50216	200	46.25
	G50218	300/500	50.25
Vaportight - Clear Globe with Guard	Appleton		
	VPOBW10G	60/200	41.00
	VPOBW20G	200/300	47.25
Explosion-Proof - Clear Globe and Guard	Appleton		
	AC1050G	100	376.00
	AC1550G	150/300	377.00
	AC2050G	200/300	511.00
	AC5050G	300/500	728.00
Add for standard reflector	(polyester)	----	54.00

The costs for these fixtures consist of the published contractors' book price for the fixtures shown. Included are the lamps indicated and labor for installation. No outlet is included. See the Branch Circuit Section for the appropriate type of outlet.

COST OF BATTERY-POWERED EMERGENCY LIGHTING UNITS

Type of Unit	Manufacturer's Catalog Number	Pro rata Guarantee	Installed Cost
Dual-Lite General Purpose	ALA-30-SB	7 year	$349.00
	ALC-X-30	15 year	388.00
	AS-145	20 year	530.00
Self-Contained Emergency Light - Dual-Lite	EDS	5 year	162.00
Semirecess Mounting Kit	F-SRM	----	27.00
Full-Recess Mounting Kit	FRM	----	31.00
Low-Voltage Fixtures	EXT-123S	----	52.00
	EXT-130	----	66.00
	EXT-P5SB	----	48.00
Self-Contained Emergency Light - Dual-Lite	Single - EDC104	5 year	140.00
	Twin - EDC204	5 year	188.00
	Single - EIC-1	10 year	198.00
	Twin - EIC-2	10 year	283.00

The costs shown for these units consist of the published contractors' book price for the units shown, which are manufactured by Dual-Lite Inc. of Newtown, Conn. The 7-year guarantee consists of what is known as a lead-acid battery, the 15-year guarantee of what is known as a lead-calcium battery, and 20-year guarantee of what is known as a nickel-cadmium battery.

COST OF BATTERY-POWERED AC SYSTEMS

Enclosure Arrangement	Battery	Emergency Power in Watts	Installed Cost
One Cabinet		500	$2,571
		750	3,043
		1000	3,142
		1250	3,700
		1500	4,666
Two Cabinets	Long-Life Lead with 10-year prorated Guarantee	2400	6,834
		3000	7,770
		3600	8.832
		4500	9,928
		5500	14,347
Three Cabinets		6500	15,201
		8000	15,971
		10000	19,044
		12000	24,200
One Cabinet		500	4,957
		750	6,454
		1000	6,623
Two Cabinets	Nickel-Cadmium Battery with 15-year pro-rated Guarantee	1250	8,090
		1500	8,563
		2400	12,084
		3000	17,015
		4500	21,033
Three Cabinets		6500	29,972
		8000	43,658
		10000	48,572
		12000	57,125

The costs shown for these units consist of the published contractors' book prices for the units illustrated as manufactured by Dual-Lite Inc. of Newtown, Conn. They include the necessary batteries to satisfy Article 700 of the NEC (87 1/2% of the normal voltage maintained for 1 1/2 hours under full-load conditions). See page A-20 for accessories.

Description	Dual-Lite Catalog Number	Capacity	Installed cost
Accessories for AC Systems			
Remote Trouble and Battery Monitor	RAP-2	----	$ 332.00
Input or Output transformers	277/120V.	750VA 1500VA 2400/3000VA 4500VA 5500/6000VA 8000VA 15000VA	196.00 378.00 465.00 585.00 1108.00 1323.00 2585.00
Self-Powered Exit Light 10-Yr. Battery	Exquisite Series	Single Face Double Face	196.00 203.00

Accessories shown include the published contractors' book price for the units and the labor for installation. Wiring between monitor unit and central equipment is not included.

The self-powered signs contain a hermetically sealed battery in a recessed box for mounting in the ceiling or wall. The fixtures have six-inch-height letters. Back boxes in this case are included as part of the fixture, as it is a special.

A-21

Type of Unit	Wattage	Installation Cost
Post Light	1-100	$77.50
Porch Light	2-60	12.25
Exterior Wall Bracket	1-100	1.50
Porch or Corridor	1-100	12.25

The cost shown for the Post Light does not include the cost of the pole or the fixture; however, it does include the cost of hand excavation, concrete base, conduit elbow, wire in the pole, and the lamp.

The costs shown for the other fixtures do not include the cost of the fixture, but do include the required lamps and the labor for the fixture installation.

Type of Unit	Wattage	Installation Cost
Chandelier	5-60	$49.75
Wall Bracket	1-60	17.25
Kitchen, etc.	2-100	12.75
Hall or Closet	1-60	12.25

The costs shown for these fixtures do not include the cost of the fixture, but do include the required lamps and the labor for the fixture installation.

A-23

Type of Unit	Wattage	Installation Cost
General Use	1-100	$25.50
Recessed	1-100	40.25
Bath	1-60	17.75
Bedroom	4-60	15.75

The costs shown for these fixtures do not include the cost of the fixture, but do include the required lamps and the labor for the fixture installation.

COST OF EXTERIOR WALL BRACKETS AND SPOTS

Type of Unit	Manufacturer's Catalog Number	Wattage & Lamp	Installed Cost
Holophane - Wallpack	WP1A &WP2A series	150W Inc 35W HPS 50W HPS 70W HPS 100W MV 175W MV	$119.00 209.00 215.00 282.00 214.00 239.00
Sylvania	AK series	50W HPS 70W HPS 100W HPS	200.00 200.00 209.00
Stonco - lampholders	150L 6700 6700S	75 PAR 150 PAR 300 R 1-150 PAR 2-150 PAR	30.00 24.00 26.00 52.00 74.00
Sylvania - Floodlight	MSK-70 MSK-100 MSK-150	70 HPS 100 HPS 150 HPS	173.00 184.00 195.00
Stonco - Power Beams	4600E 5600E 6400E Q6400E	200 PAR46 300 PAR56 500 PAR64 1000 PAR64qtz	119.00 128.00 164.00 342.00

The costs shown for these fixtures consist of the published contractors' book price for the fixtures shown. Included are the required lamps and the labor for fixture installation; however, no outlet is included. See Branch Circuit Section for the type required.

COST OF STEEL FLOODLIGHTING POLES

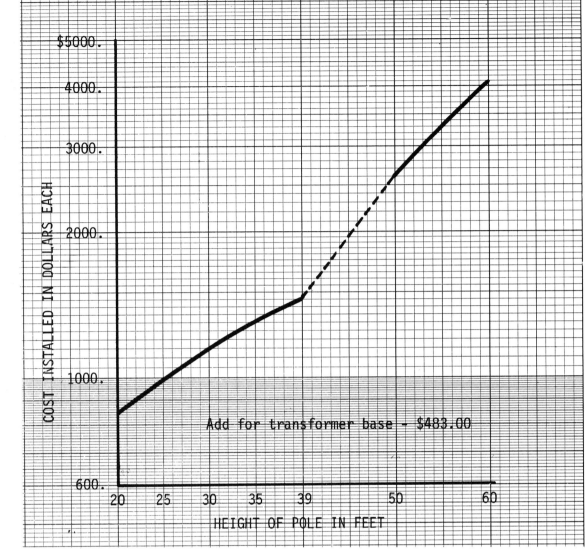

Add for transformer base - $483.00

The costs shown for these poles consist of the published contractors'
book prices with a shoe base and are as manufactured by Valmont
Industries, Inc. For a transformer base make an adder as indicated
on the graph.

Included: (material and labor)

1. Cost of the pole including freight
2. Hand excavation time in normal soil
3. Cost of concrete installed
4. Anchor bolts
5. 1" GRC elbow
6. 10' of #8 bare copper ground wire
7. A hand hole in the base of the pole
8. Grounding stud
9. Wire from the base of the pole to the luminaire

Excluded:

1. Arms, fittings, and luminaires for the top of the pole

FLOODLIGHTING POLES - ROUND TAPERED PRIMED STEEL

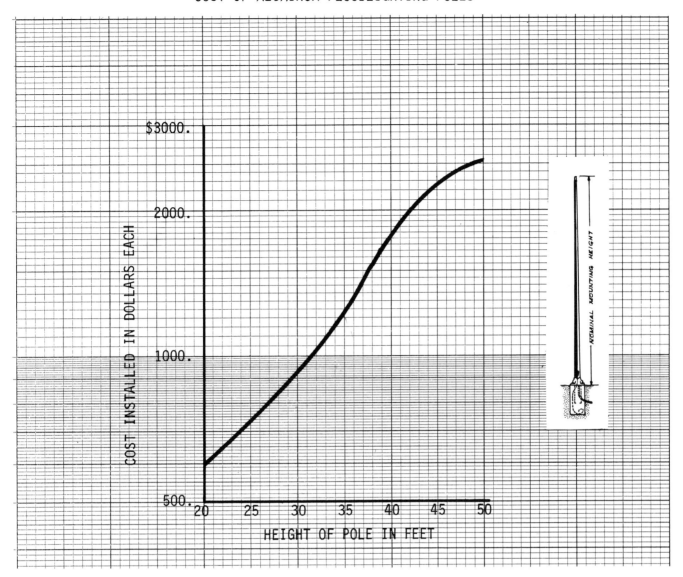

The costs shown for these poles consist of the published contractors'
book prices with a shoe base and are as manufactured by Lexington,
Inc.

Included: (material and labor)

1. Cost of the pole including freight
2. Hand excavation time in normal soil
3. Cost of concrete installed
4. Anchor bolts
5. 1" GRC elbow
6. 10' of #8 bare copper ground wire
7. A hand hole in the base of the pole
8. Grounding stud
9. Wire from the base of the pole to the luminaire

Excluded:

1. Arms, fittings, and luminaires for the top of the pole

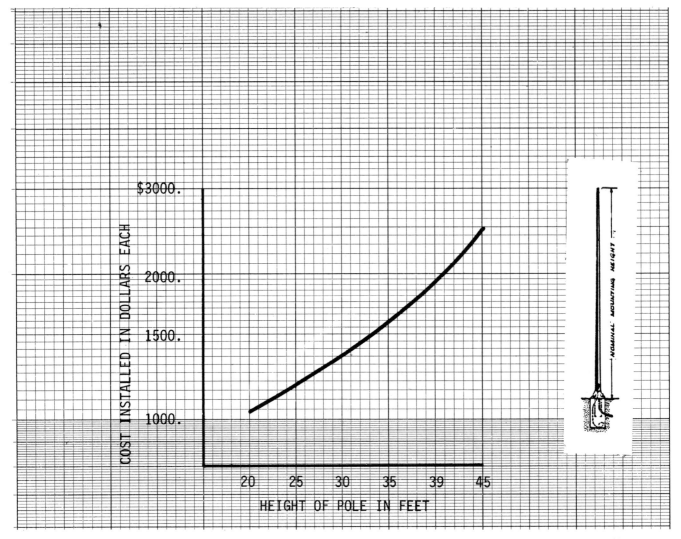

A-27

The costs shown for these poles consist of the published contractors' book prices with a shoe base and are as manufactured by Valmont Industries Inc. For a transformer base make an adder as indicated on page A-25.

Included: (material and labor)

1. Cost of the pole including freight
2. Hand excavation time in normàl soil
3. Cost of concrete installed
4. Anchor bolts
5. 1" GRC elbow
6. 10' of #8 bare copper ground wire
7. A hand hole in the base of the pole
8. Grounding stud
9. Wire from the base of the pole to the luminaire

Excluded:

1. Arms, fittings, and luminaires for the top of the pole

FLOODLIGHTING POLES - SQUARE TAPERED PRIMED STEEL

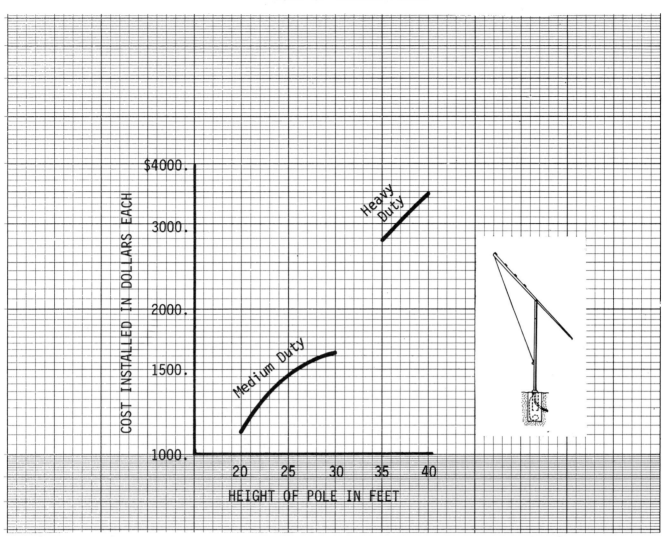

The costs shown for these hinged poles consist of published contractors'
book prices with a shoe base and are as manufactured by Valmont
Industries Inc. For a transformer base add as indicated on page A-25.

Included: (material and labor)

1. Cost of the pole including freight
2. Hand excavation time in normal soil
3. Cost of concrete installed
4. Anchor bolts
5. 1" GRC elbow
6. 10' of #8 bare copper ground wire
7. A hand hole in the base of the pole
8. Wire from the base of the pole to the luminaire

Excluded:

1. Arms, fittings, and luminaires for the top of the pole

A-29

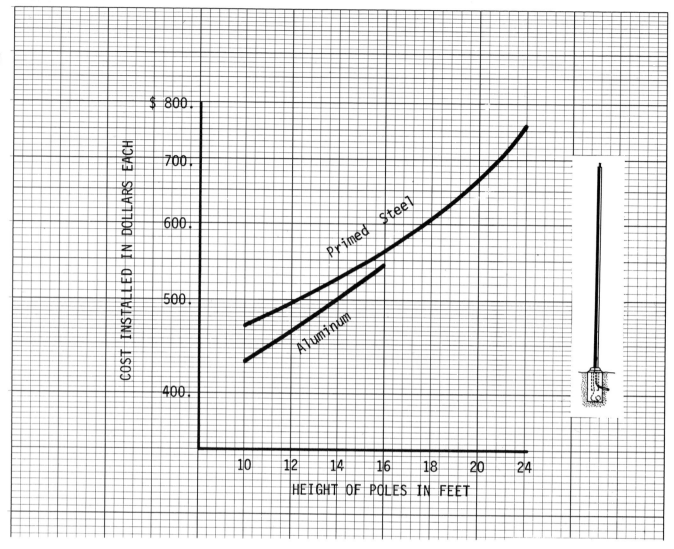

The costs shown for these poles consist of the published contractors'
book prices with a shoe base and are as manufactured by Lexington Inc.
and Valmont Industries Inc.

Included: (material and labor)

1. Cost of the pole including freight
2. Hand excavation time in normal soil
3. Cost of concrete installed
4. Anchor bolts
5. 1″ GRC elbow
6. 10′ of #8 bare copper ground wire
7. A hand hole in the base of the pole
8. Grounding stud
9. Wire from the base of the pole to the luminaire

Excluded:

1. Arms, fittings, and luminaires for the top of the pole

LIGHT-DUTY POLES - ROUND TAPERED ALUMINUM AND PRIMED STEEL

COST OF STREET-LIGHTING POLES WITH PIPE ARMS

COST INSTALLED IN DOLLARS EACH

Luminaire Mounting Height in Feet		ARM LENGTH IN FEET		
		4	6	8
GALVANIZED STEEL	21	$1004	$1024	$1045
	26	1152	1172	1191
	30	1222	1242	1262
	35	1443	1479	1499
	40	1673	1693	1712
	Add for Second Arm	132	136	148
ALUMINUM	20	591	596	642
	25	703	761	809
	30	872	947	991
	40	1402	1406	1450
	Add for Second Arm	132	136	148

The costs shown consist of the published contractors' book prices for the poles described. These poles are manufactured by Lexington Inc. and Valmont Industries Inc.

Included: (material and labor)

1. *Cost of the pole and arm including freight*
2. *Hand excavation time in normal soil*
3. *Cost of concrete installed*
4. *Anchor bolts*
5. *1" GRC elbow*
6. *10' of #8 bare copper ground wire*
7. *A hand hole in the base of the pole*
8. *Grounding stud*
9. *Wire from the base of the pole to the luminaire*

Excluded:

1. *Luminaires*

ROUND TAPERED STREET-LIGHTING POLES WITH PIPE ARMS

A-31

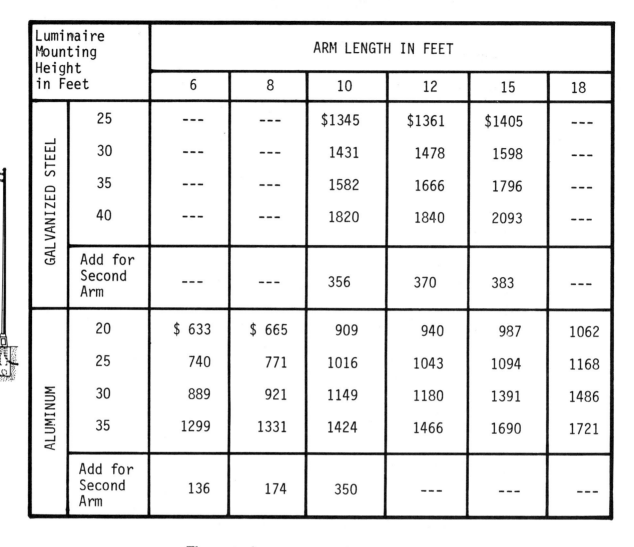

Luminaire Mounting Height in Feet		ARM LENGTH IN FEET					
		6	8	10	12	15	18
GALVANIZED STEEL	25	---	---	$1345	$1361	$1405	---
	30	---	---	1431	1478	1598	---
	35	---	---	1582	1666	1796	---
	40	---	---	1820	1840	2093	---
	Add for Second Arm	---	---	356	370	383	---
ALUMINUM	20	$ 633	$ 665	909	940	987	1062
	25	740	771	1016	1043	1094	1168
	30	889	921	1149	1180	1391	1486
	35	1299	1331	1424	1466	1690	1721
	Add for Second Arm	136	174	350	---	---	---

The costs shown consist of the published contractors' book price for the poles described. These poles manufactured by Valmont Industries Inc. and Lexington Inc.

Included: (material and labor)

1. *Cost of the pole and truss arm including freight*
2. *Hand excavation time in normal soil*
3. *Cost of concrete installed*
4. *Anchor bolts*
5. *1" GRC elbow*
6. *10' of #8 bare copper ground wire*
7. *A hand hole in the base of the pole*
8. *Grounding stud*
9. *Wire from the base of the pole to the luminaire*

Excluded:

1. *Luminaires*

ROUND TAPERED STREET-LIGHTING POLES WITH TRUSS ARMS

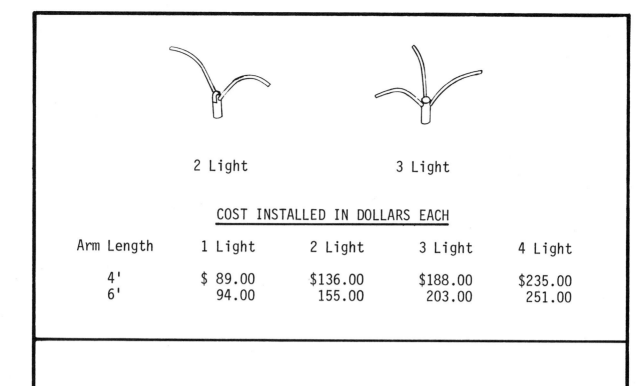

2 Light   3 Light

COST INSTALLED IN DOLLARS EACH

Arm Length	1 Light	2 Light	3 Light	4 Light
4'	$ 89.00	$136.00	$188.00	$235.00
6'	94.00	155.00	203.00	251.00

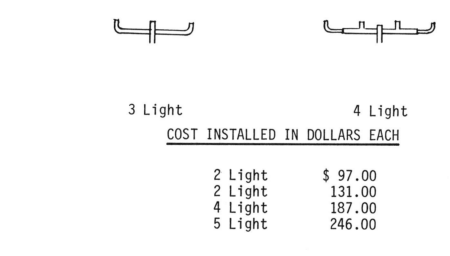

3 Light   4 Light

COST INSTALLED IN DOLLARS EACH

2 Light	$ 97.00
2 Light	131.00
4 Light	187.00
5 Light	246.00

*The costs shown for these brackets consist of the published contractors'
book prices for the brackets shown and are manufactured by Valmont
Industries Inc.*

Included: (material and labor)

  1. *Cost of the bracket including freight*
  2. *Fastening the bracket to the pole*

Excluded:

  1. *Cost of luminaire*

The costs shown for these luminaires consist of the published contractors' book prices for the luminaires illustrated.

Included: (material and labor)

    1. Luminaire with slip fitter, if required
    2. Lamp of size and type required
    3. Ballast of size required, if necessary
    4. Mounting and connecting

Excluded:

    1. Poles and brackets

Type of Unit	Wattage	Installed Cost		
		Mercury Vapor	Metal Halide	High–Pressure Sodium
Crouse-Hinds GAL Series	100	----	----	$ 592.
	150	----	----	625.
	175	$ 574.	$ 644.	----
	250	601.	----	780.
	400	747.	785.	833.
	1000	1066.	976.	1105.
Crouse-Hinds MVD Series	100	----	----	516.
	150	----	----	528.
	175	458.	520.	----
	250	505.	529.	558.
	400	502.	546.	602.
	1000	695.	760.	883.
Appleton Gen. Purp. Sportslighter	400	274.	300.	415.
	1000	----	375.	473.
	1500	----	416.	----
Appleton Quartzlite				Tungsten Halogen
	300	----	----	102.
	500	----	----	86.
	1000	----	----	124.
	1500	----	----	116.

COST OF EXTERIOR LUMINAIRES

A-34

Type of Unit	Wattage	Installed Cost		
		Mercury Vapor	Metal Halide	High-Pressure Sodium
Spaulding Peachtree Series	70	-----	-----	$399.
	100	$352.	-----	405.
	150	-----	-----	408.
	175	345.	-----	---
	250	366.	-----	---
	400	368.	-----	---
Spaulding Baltimore Series	400	557.	$607.	676
	1000	617.	690.	---
Spaulding Vega Series		800MA	1500MA	
	6'-3 lamp	-----	550.	
	6'-4 lamp	-----	633.	
	8'-4 lamp	-----	643.	
	8'-6 lamp	-----	747.	
Spaulding Fluorescent Facade Light	4'	234.	265.	
	6'	258.	280.	
	8'	283.	308.	

The costs shown for these luminaires consist of the published contractors' book prices for the luminaires illustrated.

Included: (material and labor)

1. *Luminaire with slip fitter, if required*
2. *Lamp of the size and type required*
3. *Ballast of size required, if necessary*
4. *Mounting and connecting*

Excluded:

1. *Poles and brackets*

EXTERIOR LUMINAIRES

Type of Unit	Wattage	Mercury Vapor	Metal Halide	High-Pressure Sodium
Holophane Square Post Top	70	----	----	$532.
	100	----	----	539.
	150	----	----	546.
	175	$578.	$595.	----
	250	601.	610.	718.
	400	634.	644.	769.
	1000	889.	917.	----
For Mansard Cover	Add $291.			
Gardco Form 10H Cutoff Unit	70	----	----	490.
	100	----	----	494.
	150	----	----	499.
	175	----	431.	---
	250	----	568.	652.
	400	----	596.	683.
	1000	----	808.	981.
Crouse-Hinds OV15 & OV25 Tudor Series	100	244.	----	296.
	150	----	----	297.
	175	240.	----	---
	250	285.	----	330.
	400	386.	----	429.

*The costs shown for these luminaires consist of the published contractors'
book prices for the luminaires illustrated.*

Included: (material and labor)

1. Luminaire with slip fitter, if required
2. Lamp of size and type required
3. Ballast of size required, if necessary
4. Mounting and connecting

Excluded:

1. Poles and brackets

A-36

Type of Unit	Manufacturer's Catalog Number	Wattage	Installed Cost
	#5959-S with Anchor Bolt Cover	3-60W Inc.	$ 676.00
	#62017	150 PAR 38	373.00
	#62028	100 M.V.	635.00
	#62027	150 PAR 38	479.00
	#62034	175W M.V.	747.00
	#5963-S	2-F15T8	494.00
	#5963-2S	2-f15T8	523.00

Add for other than standard gray color: $40.00

The costs shown consist of the published contractors' book prices, and lamps indicated. Direct-buried units include excavation and compaction around unit. Anchor-bolt units include hand excavation, concrete base, and installation of anchor bolts.

Manufacturer's catalog numbers are Prescolite.

Circuit Protection

BOLTED PRESSURE SWITCHES

Full-Load Break, 480/600 Volt B-1

DISTRIBUTION PANELBOARDS

Cabinet Only—Fusible Type—240/600 Volt—Main Lugs Only B-2
Cabinet and Main Switch Only—240 Volt B-3
Cabinet and Main Switch Only—600 Volt' B-4
Switch and Fuse Branches B-5
Cabinet—Circuit-Breaker Type—250/600 Volt—Main Lugs Only B-6
Cabinet and Main Breaker—250/600 Volt B-7
Circuit-Breaker Branches B-8
Cabinet Cable Interconnect B-9

BRANCH CIRCUIT PANELBOARDS

Single-Phase—120/240 Volt—3 Wire

Residential, Commercial Bolt-in and Plug-in B-10

Three-Phase—120/208 Volt—4 Wire

Residential, Commercial Bolt-in and Plug-in B-11

Circuit Breakers Only—For Panelboards B-12
Three-Phase—277/480 Volt—4 Wire
Commercial Bolt-in or Plug-in B-13

SAFETY SWITCHES

Single-Phase—240 Volt

General and Heavy Duty—Fusible B-14

Three-Phase—240 Volt—3 Wire

General and Heavy Duty—Fusible B-15

Three-Phase—240 Volt—4 Wire
General and Heavy Duty—Fusible B-16

Three-Phase—480 Volt—4 Wire

Heavy Duty—Fusible B-17

Three-Phase—600 Volt—3 Wire

Heavy Duty—Fusible B-18

Three-Phase—Nonfusible—250/600 Volt

General and Heavy Duty—General Purpose and Raintight B-19

ENCLOSED CIRCUIT BREAKERS

General Use—15-60 Amp—240, 480, & 600 Volt B-20

General Use—70-100 Amp—240, 480, & 600 Volt B-21

General Use—125-225 Amp—600 Volt B-22

General Use—250-400 Amp—600 Volt B-23

General Use—450-600 Amp—600 Volt B-24

General Use—700-800 Amp—600 Volt B-25

General Use—900-1000 Amp—600 Volt B-26

ENCLOSED MOLDED-CASE SWITCHES

General Use—100 Amp—240, 480, & 600 Volt B-27

General Use—225-1000 Amp—600 Volt B-28

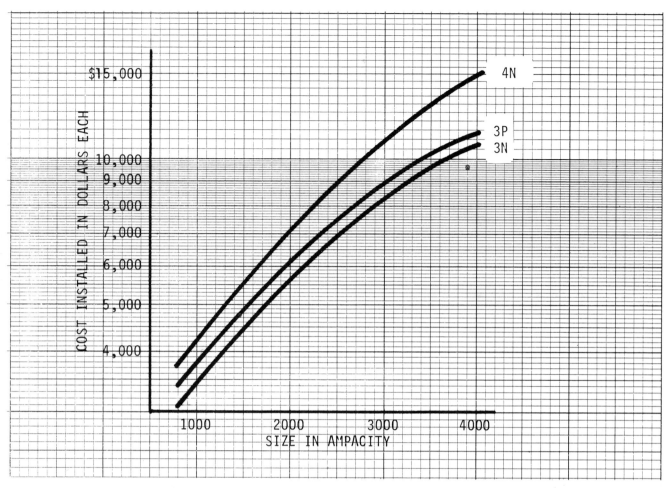

The costs shown for these bolted pressure switches consist of the published contractors' book price. The switches used here are Square D Bolt-Loc, contained in NEMA 1 enclosure, and include Bus Hi-Cap fuses to match the switch. Also included is a necessary amount of copper conductor sized to match the switch, by paralleling with wire terminals and connecting to the bus. Labor includes placing and connecting the switch and installing fuses.

ADDERS

1. Blown-main-fuse detector	$ 315.00
2. Raintight enclosure	
800—2500 amp	272.00
3000—4000 amp	333.00
3. Electric trip mechanism	652.00
4. Ground fault protection (includes sensor, solid-state relay, control transformer, wired for electric trip mechanism):	
800—2000 amp	2770.00
2500—4000 amp	2835.00

COST OF FUSIBLE-TYPE DISTRIBUTION PANELS

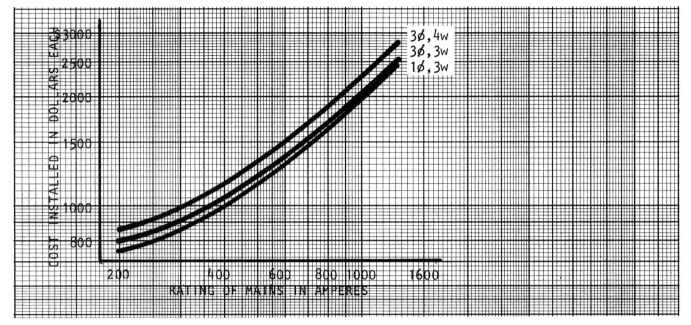

In order to determine the price of a switch or a circuit-breaker type of distribution panel, it is necessary to determine from the one-line diagram the type of cabinet, which has either main lugs, main switch, or main breaker, to which will be added the costs of the cabinet cable interconnect and the costs of the individual switches or circuit breakers.

The cost of the cabinet shown above include the following:

1. Cost of cabinet with main lugs and labor for installation.
2. Terminal time and conductor for the main connections.
3. Neutral time.
4. A grounding bar.

Not included:

1. Cable interconnect to next section.
2. Individual switches or breakers.
3. Cost of cutting holes in steel panels.

	No. of Poles	ADDERS				
		AMPACITY OF PANEL MAINS				
		225	400	600	800	1200
Subfeed Lugs	2	$ 34.00	$ 65.00	$129.00	$180.00	$215.00
	3	47.00	78.00	141.00	189.00	238.00
Split Bus	2	92.00	109.00	129.00	180.50	216.50
	3	109.50	129.00	141.00	203.00	236.00

FUSIBLE-TYPE PANELBOARD - CABINET WITH MAIN LUGS ONLY 240/600 VOLT

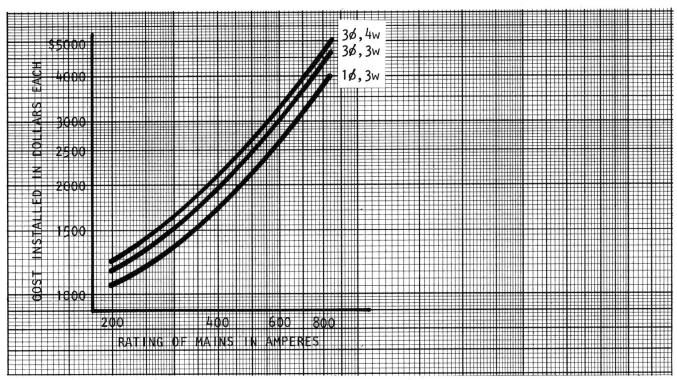

In order to determine the price of a switch or a circuit-breaker type of distribution panel, it is necessary to determine from the one-line diagram the type of cabinet, which has either main lugs, main switch, or a main breaker, to which will be added the costs of the cabinet cable interconnect and the costs of the individual switches or circuit breakers.

The cost of the cabinet shown on this graph includes the following:

1. Cost of cabinet with switch and fuses as selected and labor for installation.
2. Terminal time and conductor for the main connections.
3. Neutral time.
4. A grounding bar.

Not included:

1. Cable interconnect to next section.
2. Individual branch circuit switches.
3. Cost of cutting holes in steel panels.

	No. of Poles	ADDERS			
		AMPERE RATING			
		225	400	600	800
Subfeed Lugs	2	$ 34.00	$ 65.00	$129.00	$180.00
	3	47.00	78.00	141.00	189.00
Split Bus	2	92.00	109.00	129.00	180.50
	3	109.50	129.00	141.00	203.00

FUSIBLE-TYPE PANELBOARD CABINET – WITH 240-VOLT MAIN SWITCH

B-4

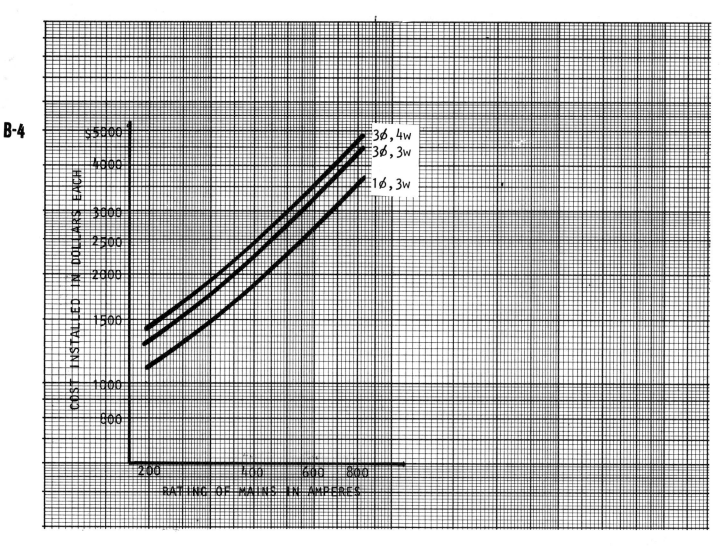

In order to determine the price of a switch or a circuit-breaker type of
distribution panel, it is necessary to determine from the one-line
diagram the type of cabinet, which has either main lugs, main switch,
or a main breaker, to which will be added the costs of the cabinet cable
interconnect and the costs of the individual switches or circuit breakers.

The cost of the cabinet shown on this graph includes the following:

1. Cost of cabinet with switch and fuses as selected and labor for
   installation.
2. Terminal time and conductor for the main connections.
3. Neutral time.
4. A grounding bar.

Not included:

1. Cable interconnect to next section.
2. Individual branch circuit switches.
3. Cost of cutting holes in steel panels.

ADDERS

For subfeed lugs or split bus, see page B-3.

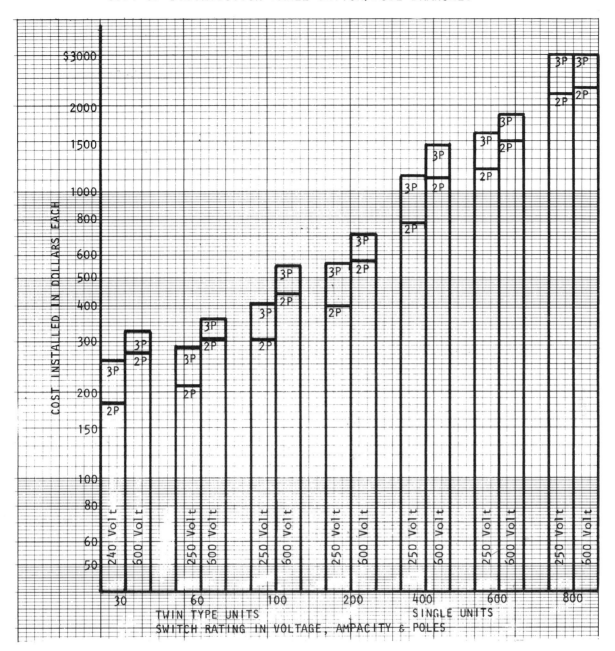

B

B-5

 After the price of the cabinet and cable interconnect cost are determined, the number of
individual switches of the proper class and number of poles must be added for the total
installed cost of the distribution panel. Each switch includes the cost of the conductors,
terminal makeup time, and time for a neutral connection, as well as cabinet time. No
cost of cutting holes in steel panels for conduit terminals has been provided. Add as
needed.

ADDER FOR 3-POLE SPACES ONLY		
AMPACITY	240 Volt	600 Volt
30-30	$ 51.00	$ 51.00
60-60	51.00	69.00
100-100	69.00	80.00
200	100.00	100.00
400	151.00	151.00
600	151.00	151.00
800	287.00	287.00

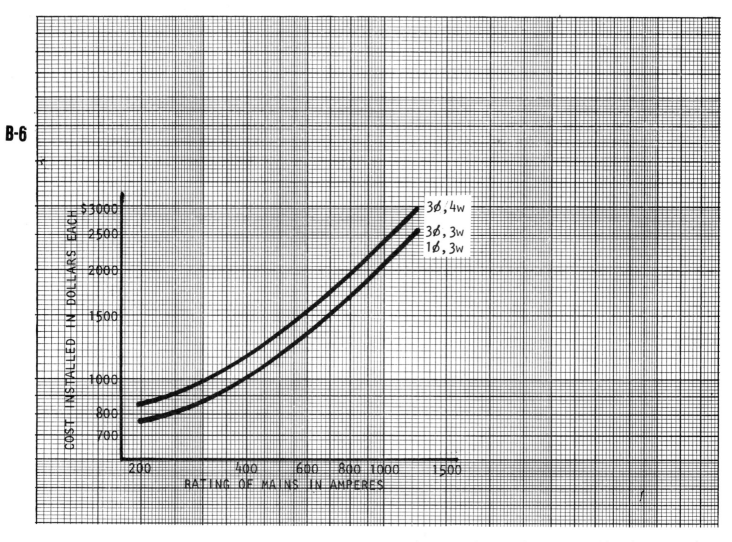

B-6

In order to determine the price of a switch or a circuit-breaker type of distribution panel, it is necessary to determine from the one-line diagram the type of cabinet, which has either main lugs, main switch, or a main breaker, to which will be added the costs of the cabinet cable interconnect and the costs of the individual switches or circuit breakers.

The cost of the cabinet shown on the above graph includes the following:

1. Cost of cabinet with main lugs and labor for installation.
2. Terminal time and conductor for the main connections.
3. Neutral time.
4. A grounding bar.

Not included:

1. Cable interconnect to next section.
2. Individual branch circuit breakers.
3. Cost of cutting holes in steel panels.

ADDERS

For subfeed lugs or split bus, see page B-3.

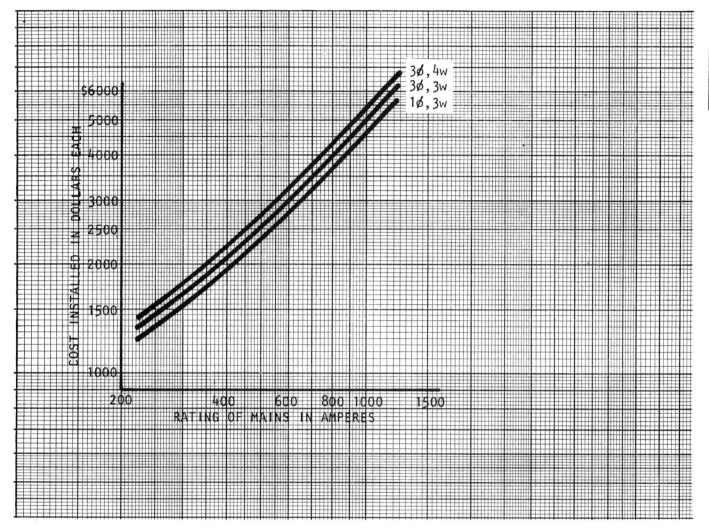

In order to determine the price of a switch or a circuit-breaker type of distribution panel, it is necessary to determine from the one-line diagram the type of cabinet, which has either main lugs, main switch, or a main breaker, to which will be added the costs of the cabinet cable interconnect and the costs of the individual switches or circuit breakers.

The cost of the cabinet shown on the above graph includes the following:

1. Cost of cabinet with main circuit breaker as selected and labor for installation.
2. Terminal time and conductor for the main connections.
3. Neutral time.
4. A grounding bar.

Not included:

1. Cable interconnect to next section.
2. Individual branch circuit breakers.
3. Cost of cutting holes in steel panels.

ADDERS

For subfeed lugs or split bus, see page B-3.

B-8

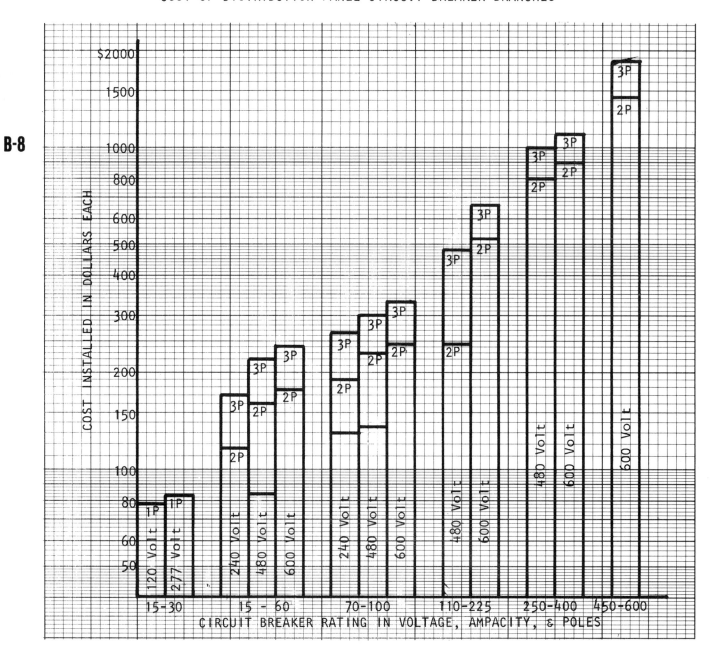

After the price of the cabinet and cable interconnect cost are determined, the number of individual switches of the proper class and number of poles must be added for the total installed cost of the distribution panel. Each circuit breaker includes the cost of the conductors, terminal makeup time, and time for a neutral connection, as well as cabinet time. No cost of cutting holes in steel panels for conduit terminals has been provided. Add as needed.

ADDER FOR SPACES ONLY			
AMPACITY	1 Pole	2 Pole	3 Pole
15–100	$12.00	$13.50	$17.50
100–225	N.A.	17.50	17.50
250–400	N.A.	38.50	38.50
450–800	N.A.	76.00	76.00

COST OF PANEL INTERCONNECT (CABLE)

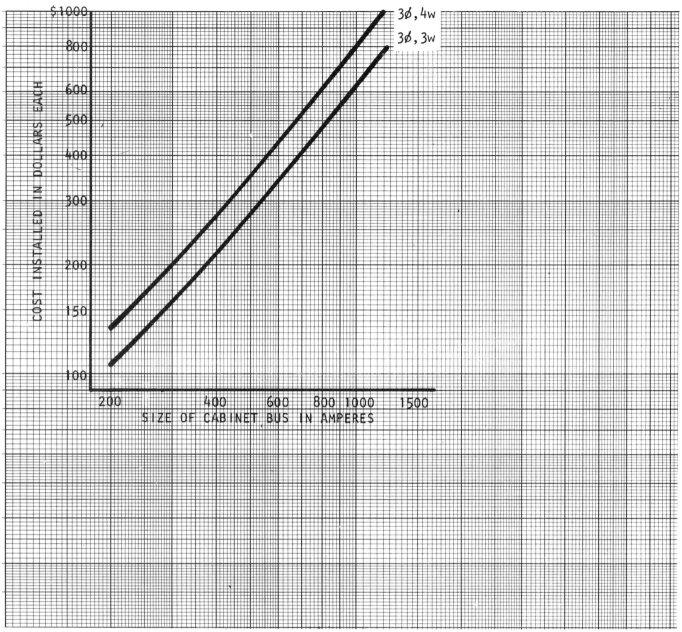

The cable interconnect consists of copper conductors for the ampacities shown, the necessary feeder conductor terminals, and the labor for installation. It is used when cross-busing is not provided by the manufacturer.

B-10

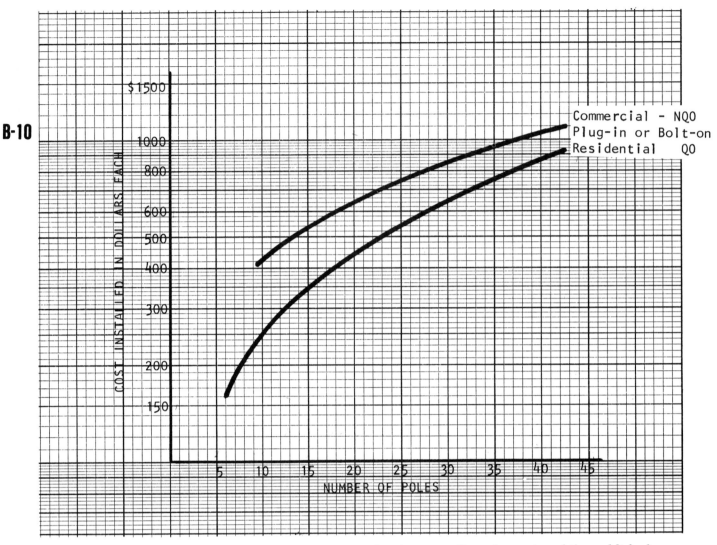

The costs shown for these panelboards consist of the published contractors' book price, and also include wire terminals for the number of breakers provided. The branch circuit conduit terminals are not included here as they are provided with the outlet boxes. A feeder conduit and feeder wire terminal, however, are provided, of a size adequate to feed the panel with the number of branch circuits provided. Also provided is sufficient feeder conductor for connections to the main buses.

The residential panel referenced would be the equivalent of the Square D QO Load Center with single-pole circuit breakers, plug-in type, rated at 10,000 A.I.C.

The commercial panels referenced would be equivalent to the Square D type NQO and NQOB, factory-assembled with single-pole plug-in or bolt-on breakers rated at 10,000 A.I.C.

ADDERS

For main breakers, see "Enclosed Breakers" in this section.

Subfeed Lugs	$19.00
Split Bus	.45.00

COST OF PANELBOARDS

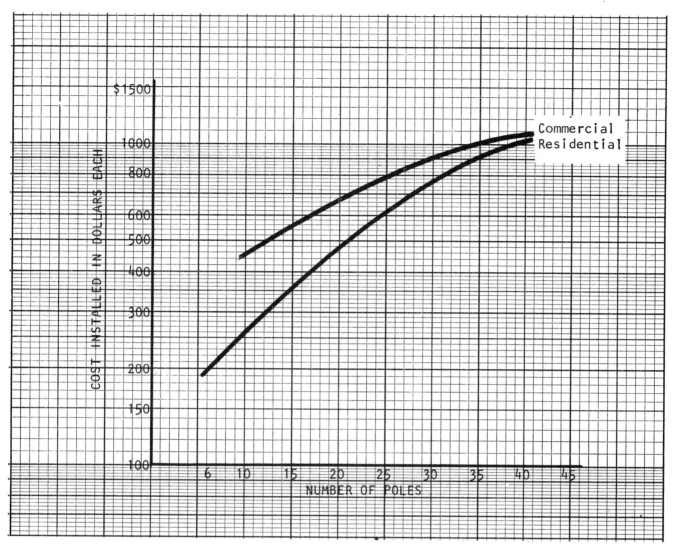

The costs shown for these panelboards consist of the published contractors' book price, and also include wire terminals for the number of breakers provided. The branch circuit conduit terminals are not included here since they are included with the outlet boxes. A feeder conduit and feeder wire terminal, however, are provided, of a size adequate to feed the panel with the number of branch circuits provided. Also provided is sufficient feeder conductor for connections to the main buses.

The residential panels referenced would be the equivalent of the Square D QO Load Center with single-pole circuit breakers, plug-in type, rated at 10,000 A.I.C.

The commercial panels referenced would be equivalent to the Square D NQO and NQOB, factory-assembled with single-pole bolt-in or plug-in breakers rated at 10,000 A.I.C.

ADDERS

For main breakers, see "Enclosed Breakers" in this section.

Subfeed Lugs	$19.00
Split Bus, 3 Pole	63.00

PANELBOARDS − 120/208, 3 PHASE, 4 WIRE − MAIN LUGS ONLY

120/240 VOLT

B-12

QO(Plug-in), QOB(Bolt-on) and QH			
10,000 A.I.C.	1 Pole	2 Pole	3 Pole
10-60A	$22.75	$36.75	$77.00
70A	29.00	49.50	90.00
80-100A	51.75	63.00	101.00
GFI 15-30A	55.50	92.25	- - -
22,000 A.I.C. 15-60A	29.25	50.50	101.50
70A	- - -	66.75	121.50
80-100A	- - -	87.75	136.25
GFI 15-30A	79.50	- - -	- - -
65,000 A.I.C. 15-30A	36.25	72.00	116.25

The costs shown for circuit breakers only consist of the published contractors' book prices and the labor for installing in an existing panelboard with spaces available. Installation of branch circuit conductor terminations is also included.

The costs shown for these panelboards consist of the published contractors' book price, and also include wire terminals for the number of breakers provided. The branch circuit conduit terminals are not included here as they are provided with the outlet box. A feeder conduit and feeder wire terminal, however, are provided, of a size adequate to feed the panel with the number of branch circuits provided. Also provided is sufficient feeder conductor for connections to the main buses.

The panels referenced would be the equivalent of the Square D type NEHB, factory-assembled with single-pole plug-in or bolt-in breakers. Single-pole breakers for this panel are rated at 14,000 A.I.C. at 277 volts.

ADDERS

For main breakers, see "Enclosed Breakers" in this section.

Subfeed Lugs	$19.00
Split Bus, 3 Pole	63.00

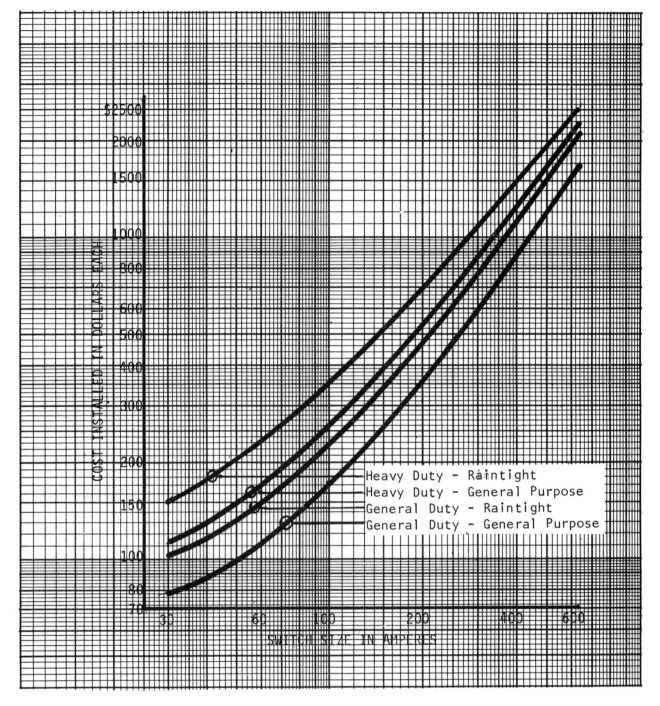

The costs shown for safety switches consist of the published contractors' book prices and further include the following:

1. *Fuses where required.*
2. *Fastening devices for mounting to a masonry wall.*
3. *A conduit terminal for the conduit size required.*
4. *Wire required inside the enclosure.*
5. *Terminations.*
6. *Labor for complete installation of switch, conduit terminal, wire, and fuses.*

SAFETY SWITCH - 240 VOLT - SINGLE PHASE - 3 WIRE

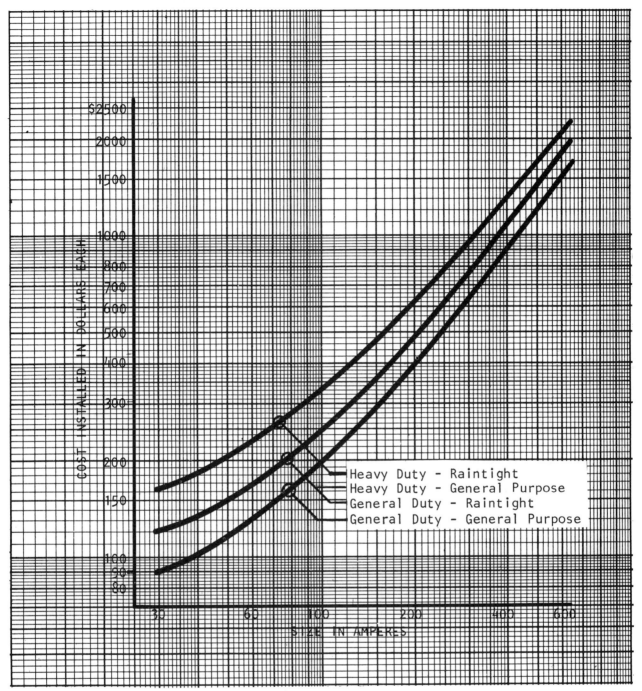

The costs shown for safety switches consist of the published
contractors' book prices and further include the following:

1. Fuses where required.
2. Fastening devices for mounting to a masonry wall.
3. A conduit terminal for the conduit size required.
4. Wire required inside the enclosure
5. Terminations.
6. Labor for complete installation of switch, conduit terminal,
    wire, and fuses.

B-16

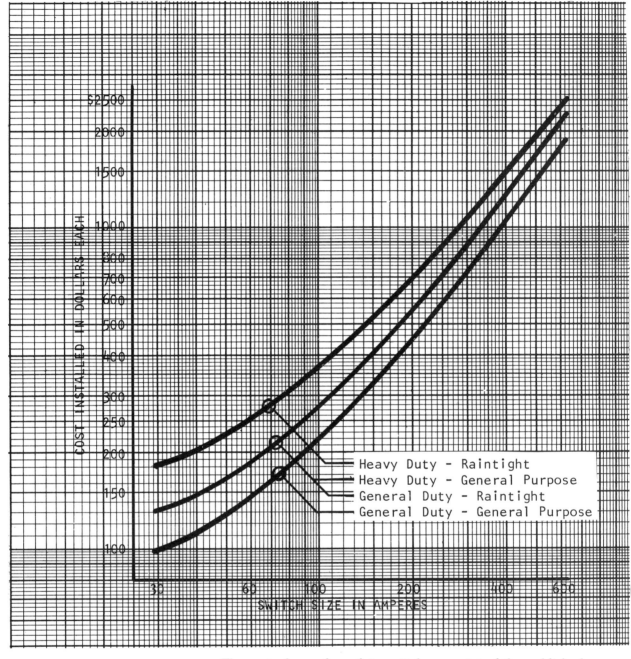

The costs shown for safety switches consist of the published contractors' book prices and further include the following:

1. Fuses where required.
2. Fastening devices for mounting to a masonry wall.
3. A conduit terminal for the conduit size required.
4. Wire required inside the enclosure.
5. Terminations.
6. Labor for complete installation of switch, conduit terminal, wire, and fuses.

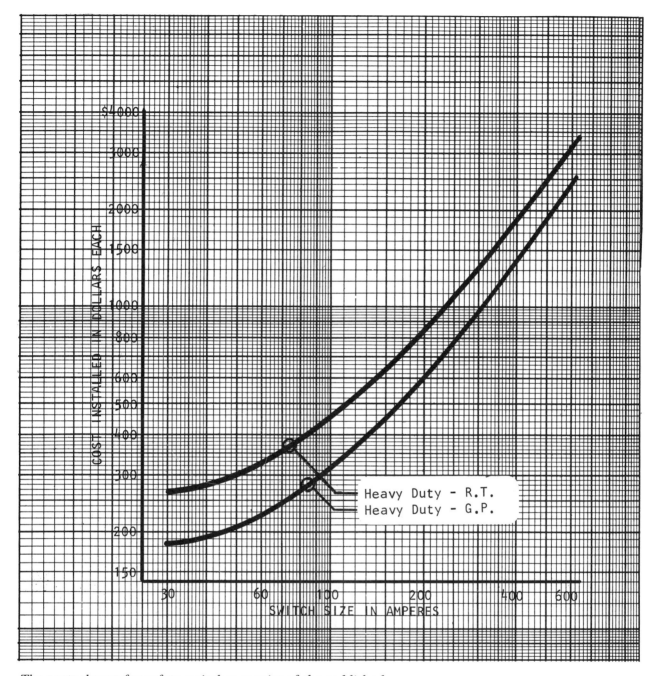

The costs shown for safety switches consist of the published contractors' book prices and further include the following:

1. Fuses where required.
2. Fastening devices for mounting to a masonry wall.
3. A conduit terminal for the conduit size required.
4. Wire required inside the enclosure.
5. Terminations.
6. Labor for complete installation of switch, conduit terminal, wire, and fuses.

B-18

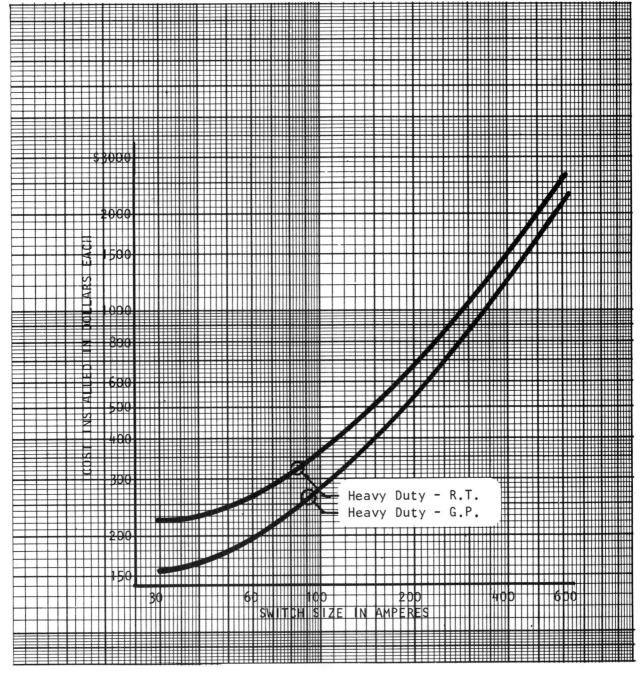

The costs shown for safety switches consist of the published contractors' book prices and further include the following:

1. Fuses where required.
2. Fastening devices for mounting to a masonry wall.
3. A conduit terminal for the conduit size required.
4. Wire required inside the enclosure.
5. Terminations.
6. Labor for complete installation of switch, conduit terminal, wire, and fuses.

FUSIBLE SAFETY SWITCH - 600 VOLT - 3 PHASE - 3 WIRE

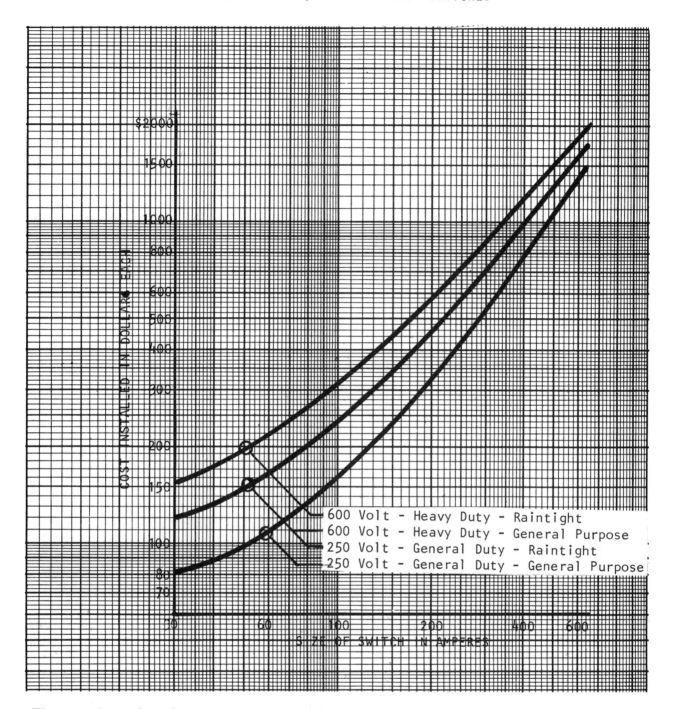

The costs shown for safety switches consist of the published contractors'
book prices and further include the following:

1. *Fastening devices for mounting to a masonry wall.*
2. *A conduit terminal for the conduit size required.*
3. *Wire required inside the enclosure.*
4. *Terminations.*
5. *Labor for complete installation of switch, conduit terminal, wire,*
   and fuses.

UL–LISTED AMPERE INTERRUPTING RATING OF SOME SQUARE D CIRCUIT BREAKERS				
TYPE OF BREAKER	120 Volt	240 Volt	480 Volt	600 Volt
QO	10,000	10,000		
QO-VH		22,000		
QH		65,000		
Q1		10,000		
Q1-VH		22,000		
EH	65,000	65,000	14,000	
FA-240	10,000	10,000		
FA-480	18,000	18,000	14,000	
FA-600		18,000	14,000	14,000
FH	65,000	65,000	25,000	18,000
KA		25,000	25,000	22,000
KH		65,000	35,000	25,000
LA		42,000	30,000	22,000
LH		65,000	35,000	25,000
MA		42,000	30,000	22,000
MH		65,000	50,000	25,000

The costs shown for the enclosed circuit breakers in general-purpose and raintight enclosures consist of the published contractors' book prices, and also include materials for fastening to a masonry wall, a conduit terminal for the size required by the wire, and wire in the amount necessary for makeup. Raintight enclosures include a conduit hub of the proper size. Labor is provided for a complete installation.

Note that the breakers are not shown in ampacity. It is assumed that the largest in its class is used in establishing an estimated price. The following references are used:

2P=2 wire-2 protected poles
3N=3 wire-2 protected poles plus solid neutral
3P=3 wire-3 protected poles
4N=4 wire-3 protected poles plus solid neutral

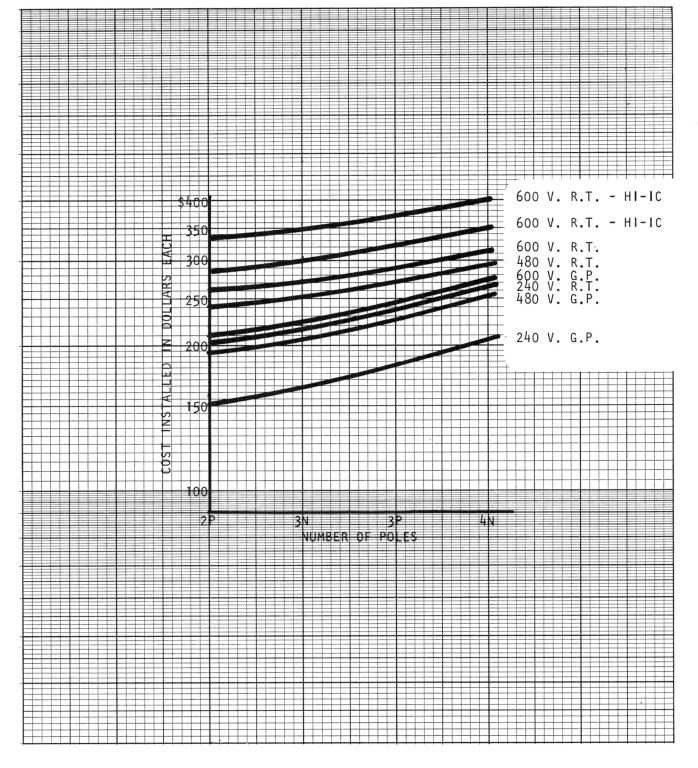

COST INSTALLED IN DOLLARS EACH

$400

350

300

250

200

150

100

2P   3N   3P   4N

NUMBER OF POLES

600 V. R.T. – HI-IC

600 V. R.T. – HI-IC

600 V. R.T.
480 V. R.T.
600 V. G.P.
240 V. R.T.
480 V. G.P.

240 V. G.P.

B-21

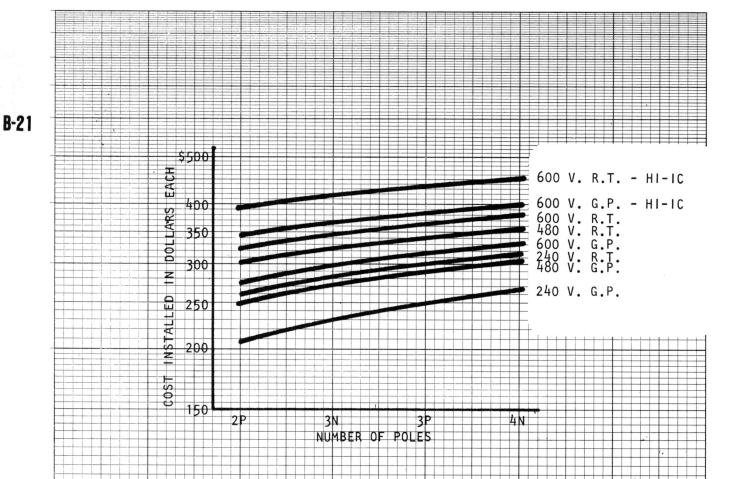

The costs shown for the enclosed circuit breakers in general-purpose and raintight enclosures consist of the published contractors' book prices, and also include materials for fastening to a masonry wall, a conduit terminal for the size required by the wire, and wire in the amount necessary for makeup. Raintight enclosures include a conduit hub of the proper size. Labor is provided for a complete installation.

Note that the breakers are not shown in ampacity. It is assumed that the largest in its class is used in establishing an estimated price. The following references are used:

  2P=2 wire-2 protected poles
  3N=3 wire-2 protected poles plus solid neutral
  3P=3 wire-3 protected poles
  4N=4 wire-3 protected poles plus solid neutral

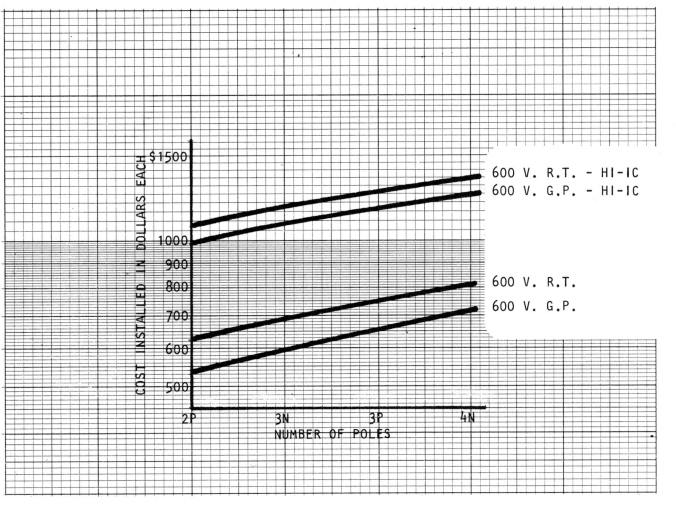

The costs shown for the enclosed circuit breakers in general-purpose and raintight enclosures consist of the published contractors' book price, and also include materials for fastening to a masonry wall, a conduit terminal for the size required by the wire, and wire in the amount necessary for makeup. Raintight enclosures include a conduit hub of the proper size. Labor is provided for a complete installation.

Note that the breakers are not shown in ampacity. It is assumed that the largest in its class is used in establishing an estimated price. The following references are used:

  2P=2 wire-2 protected poles
  3N=3 wire-2 protected poles plus solid neutral
  3P=3 wire-3 protected poles
  4N=4 wire-3 protected poles plus solid neutral

B-23

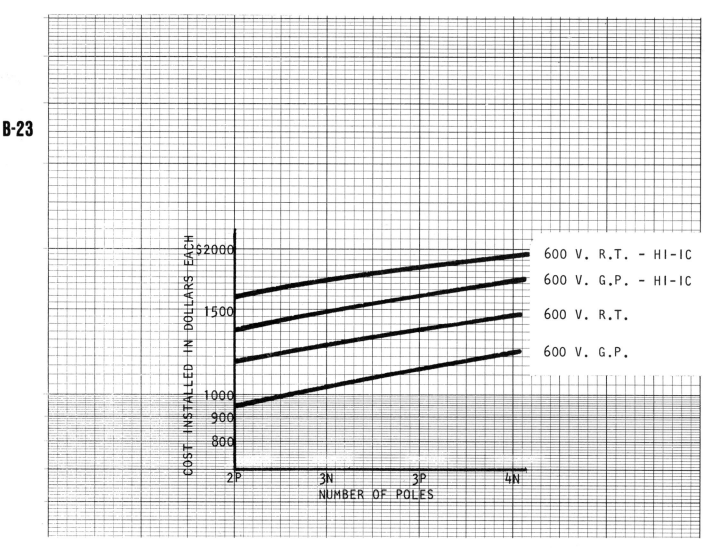

The costs shown for the enclosed circuit breakers in general-purpose and raintight enclosures consist of the published contractors' book price, and also include materials for fastening to a masonry wall, a conduit terminal for the size required by the wire, and wire in the amount necessary for makeup. Raintight enclosures include a conduit hub of the proper size. Labor is provided for a complete installation.

Note that the breakers are not shown in ampacity. It is assumed that the largest in its class is used in establishing an estimated price. The following references are used:

    2P=2 wire-2 protected poles
    3N=3 wire-2 protected poles plus solid neutral
    3P=3 wire-3 protected poles
    4N=4 wire-3 protected poles plus solid neutral

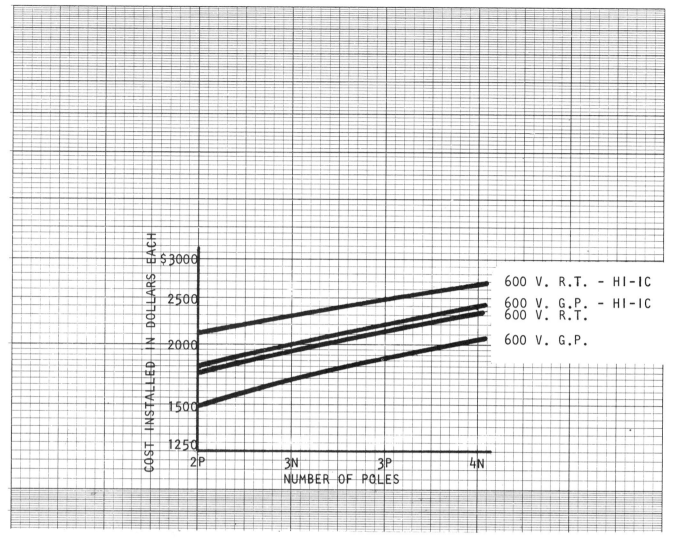

The costs shown for the enclosed circuit breakers in general-purpose and raintight enclosures consist of the published contractors' book price, and also include materials for fastening to a masonry wall, a conduit terminal for the size required by the wire, and wire in the amount necessary for makeup. Raintight enclosures include a conduit hub of the proper size. Labor is provided for a complete installation.

Note that the breakers are not shown in ampacity. It is assumed that the largest in its class is used in establishing an estimated price. The following references are used:

2P=2 wire-2 protected poles
3N=3 wire-2 protected poles plus solid neutral
3P=3 wire-3 protected poles
4N=4 wire-3 protected poles plus solid neutral

B-25

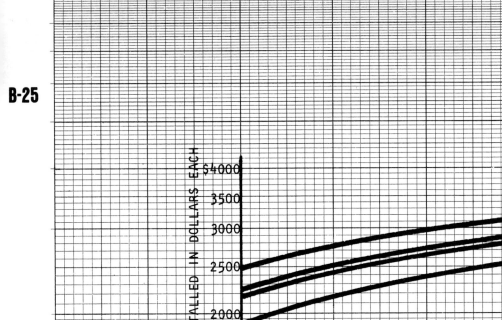

The costs shown for the enclosed circuit breakers in general-purpose and raintight enclosures consist of the published contractors' book price, and also include materials for fastening to a masonry wall, a conduit terminal for the size required by the wire, and wire in the amount necessary for makeup. Raintight enclosures include a conduit hub of the proper size. Labor is provided for a complete installation.

Note that the breakers are not shown in ampacity. It is assumed that the largest in its class is used in establishing an estimated price. The following references are used:

  2P=2 wire-2 protected poles
  3N=3 wire-2 protected poles plus solid neutral
  3P=3 wire-3 protected poles
  4N=4 wire-3 protected poles plus solid neutral

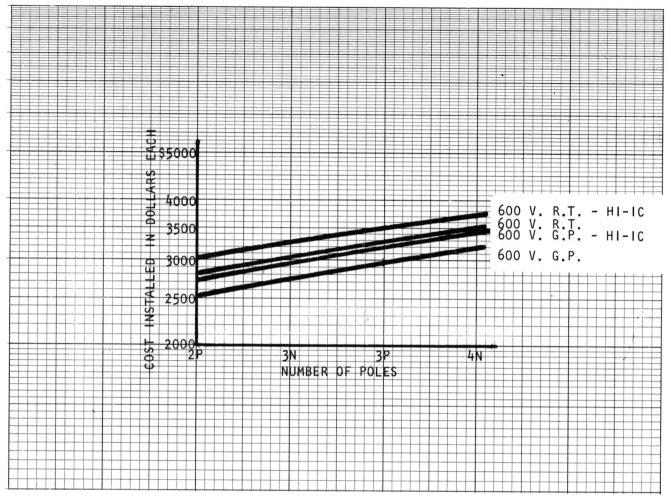

The costs shown for the enclosed circuit breakers in general-purpose and raintight enclosures consist of the published contractors' book prices, and also include materials for fastening to a masonry wall, a conduit terminal for the size required by the wire, and wire in the amount necessary for makeup. Raintight enclosures include a conduit hub of the proper size. Labor is provided for a complete installation.

Note that the breakers are not shown in ampacity. It is assumed that the largest in its class is used in establishing an estimated price. The following references are used:

2P=2 wire-2 protected poles
3N=3 wire-2 protected poles plus solid neutral
3P=3 wire-3 protected poles
4N=4 wire-3 protected poles plus solid neutral

B-27

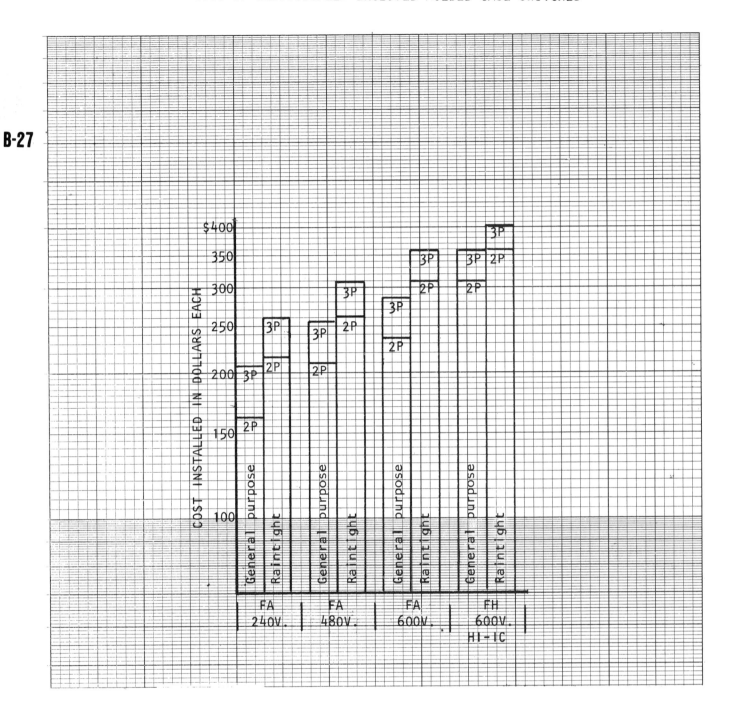

The costs shown for the enclosed molded-case switches in general-purpose and raintight enclosures consist of the published contractors' book prices, and also include materials for fastening to a masonry wall, a conduit terminal of the size required by the wire, and wire in the amount necessary for makeup. Raintight enclosures include a conduit hub of the proper size. Labor is provided for a complete installation.

100 Amp MOLDED-CASE SWITCH - SEPARATELY ENCLOSED 240, 480, 600V FA & FH

COST OF INDIVIDUALLY ENCLOSED MOLDED-CASE SWITCHES

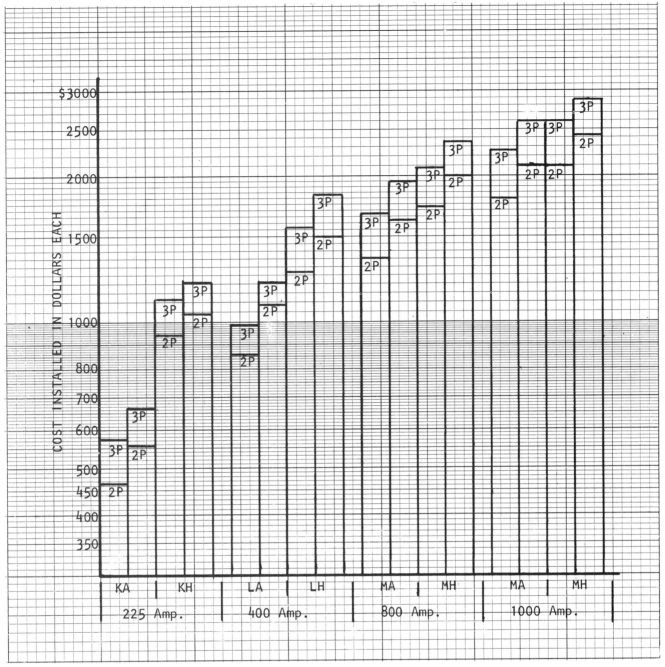

The costs shown for the enclosed molded-case switches in general-purpose and raintight enclosures consist of the published contractors' book prices, and also include materials for fastening to a masonry wall, a conduit terminal of the size required by the wire, and wire in the amount necessary for makeup. Raintight enclosures include a conduit hub of the proper size. Labor is provided for a complete installation.

225-1000 Amp MOLDED-CASE SWITCH - SEPARATELY ENCLOSED 600V KA, KH, LA, LH, MA & MH

Services

C

SERVICE ENTRANCE

Service Entrance Weatherhead   C-1

CURRENT TRANSFORMER CABINETS

Single and Three Phase—Full Neutral   C-2

MULTIMETERING

Service Main Disconnect—Switch Type   C-3
Service Main Disconnect—Circuit-Breaker Type   C-4
Group Meter Sockets: 100/125 and 200 Amp—Single Phase   C-5, C-6
Group Meter Sockets: 100/125 and 200 Amp—Three Phase   C-7

SERVICE GUTTER OR WIREWAY

Wireway—Screw or Hinged Cover—NEMA 1   C-8
Wireway Fittings—NEMA 1—Tees and Elbows   C-9
Wireway—Raintight—NEMA 3R   C-10

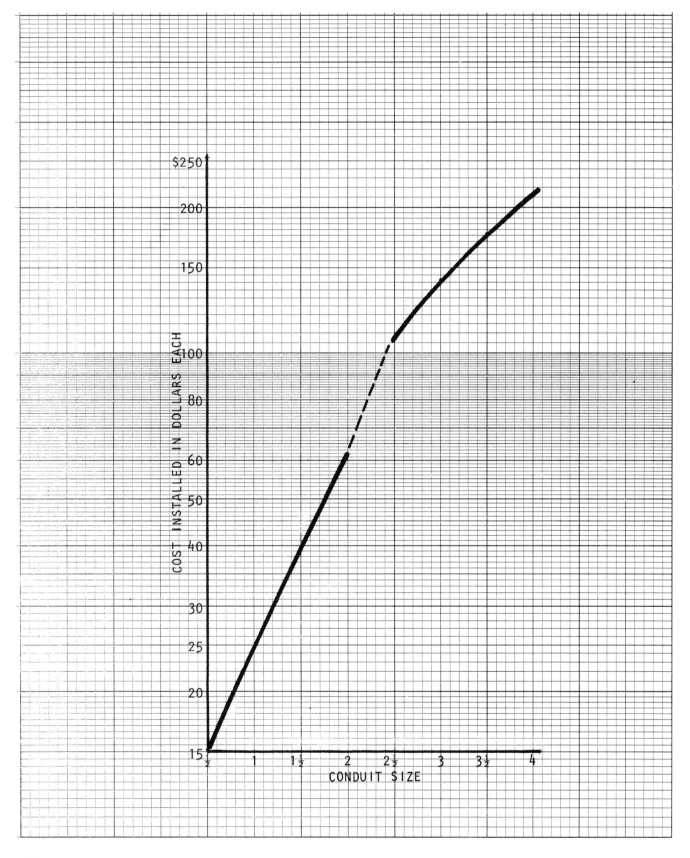

The costs shown for the service entrance weatherhead include the
published contractors' book price and the labor for preparing the
conduit for attaching. The overall installed cost for either EMT or GRC
fittings is too insignificant to plot separate curves.

C-2

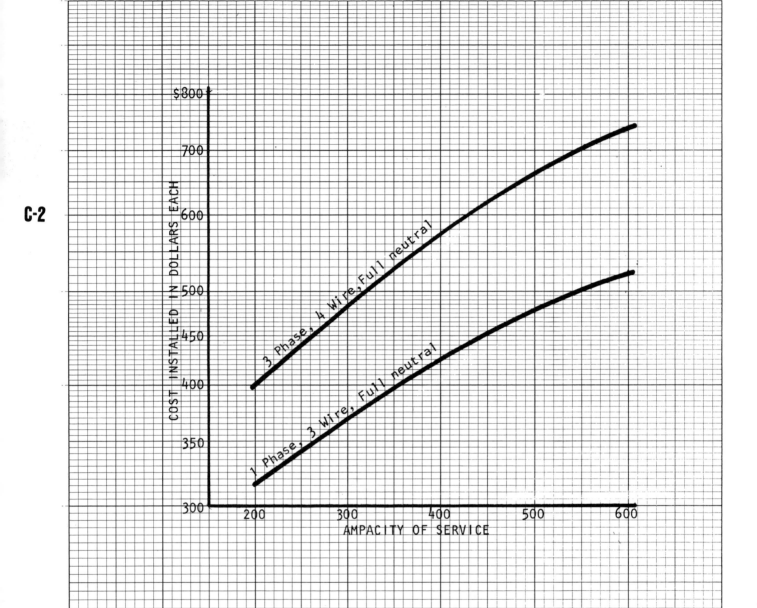

The costs shown for the current transformer cabinets consist of the contractors' book price for a cabinet 30" × 24" × 9 1/2". Also included are a neutral bus, fastening devices for mounting to a masonry wall, and conductors of the number and size required. It is assumed that the transformers are furnished by the utility company and are installed by the contractor.

Labor is included for installing the cabinet and transformers, making up the conductor ends and connecting to the transformers, and cutting two holes in the cabinet for conduit terminals; however, the conduit terminals are not provided.. See Section E for nipples and conduit terminals.

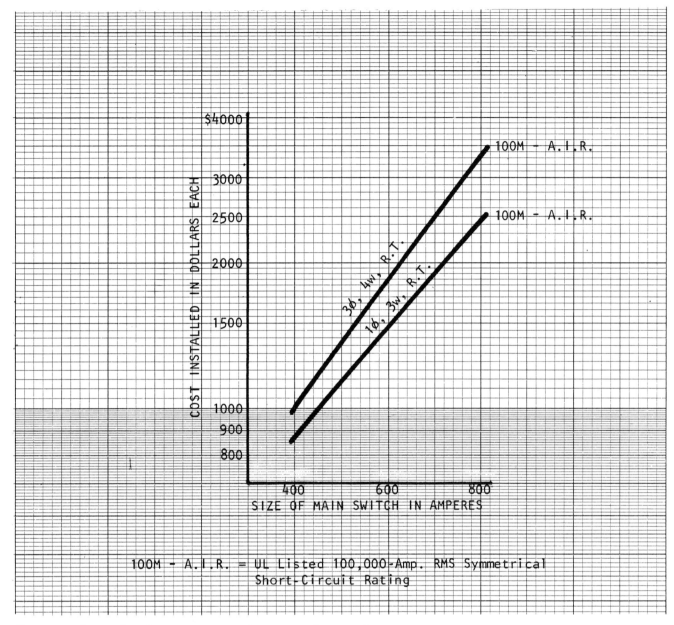

C
C-3

100M - A.I.R. = UL Listed 100,000-Amp. RMS Symmetrical Short-Circuit Rating

The costs shown for the main service disconnect for a multimetering center are based on the published contractors' book prices and are as manufactured by Square D Co. for outdoor installation. The main disconnect shown is a section containing a fused disconnect switch in single- or three-phase. The costs shown are representative of the total installed costs.

Included: (material and labor)
1. Cabinet of the type required.
2. Fuses.
3. Fastening devices for surface mounting to masonry wall.
4. Conduit terminals of size and proper number to suit ampacity.
5. Sufficient conductors of length and size to match ampacity and terminal makeup.

Excluded:
1. Raintight hubs for service conduit, as an underground supply is assumed.

The costs shown for the main service disconnect for a multimetering center are based on the published contractors' book prices and are as manufactured by Square D Co. for outdoor installation. The main disconnect shown is a section containing a circuit breaker in single- or three-phase. The costs shown are representative of the total installed costs.

Included: (material and labor)

1. Cabinet of the type required.
2. Circuit breaker of type required.
3. Fastening devices for surface mounting to masonry wall.
4. Conduit terminals of size and proper number to suit ampacity.
5. Sufficient conductors of length and size to match ampacity and terminal makeup.

Excluded:

1. Raintight hubs for service conduit, as an underground supply is assumed.

COST OF MULTIMETERING SERVICE DISCONNECT CIRCUIT BREAKER

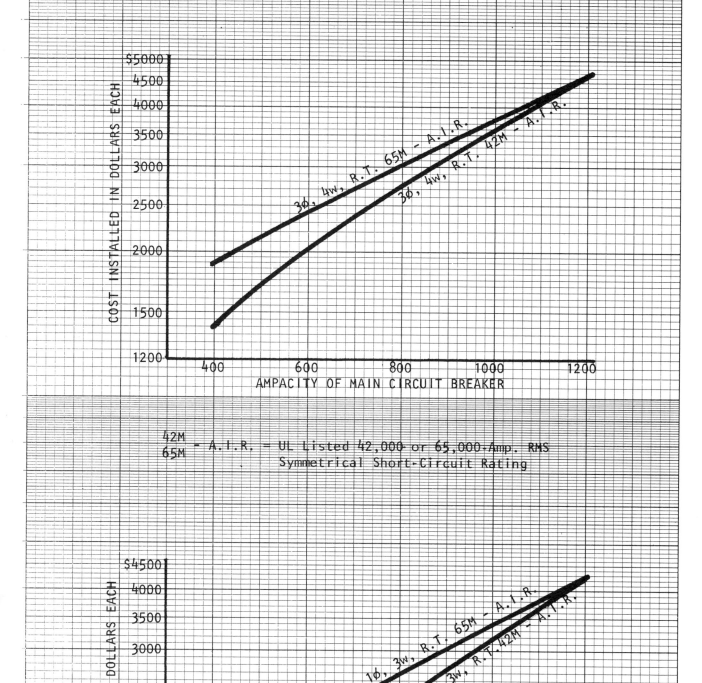

$\frac{42M}{65M}$ - A.I.R. = UL Listed 42,000 or 65,000-Amp. RMS Symmetrical Short-Circuit Rating

C-5

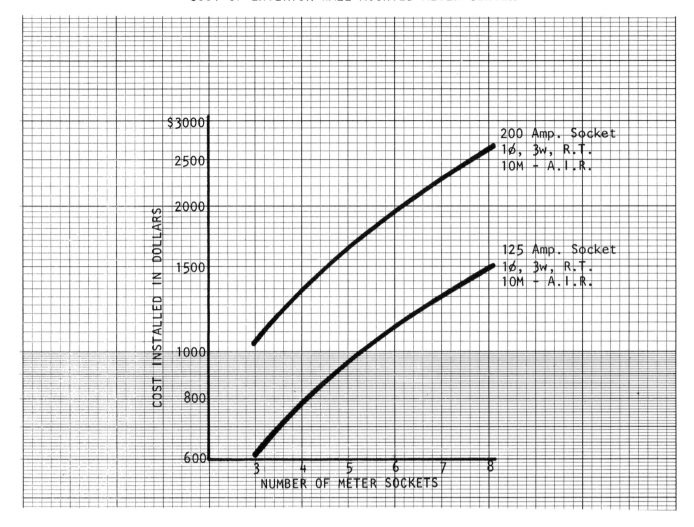

The costs shown for the wall-mounted meter center are based on the published contractors' book prices and are as manufactured by Square D Co. as an outdoor type of meter module.

These meter sockets are single-phase and shown in 100/125-ampere and 200- ampere sizes, the larger generally used for electrically heated apartments. The total installed cost difference between indoor and raintight meter sockets is not significant enough to justify two curves.

Note: While the mains may be single- or three-phase, all services to individual occupancies are single-phase.

Included: (material and labor)
1. Socket type of group and size required.
2. Two pole circuit breakers for each feeder.
3. Fastening devices for fastening to masonry wall.
4. GRC conduit terminals for each meter in the group.
5. Sufficient conductors of length and size to match ampacity and terminal makeup for each feeder.
6. Connection of main buses.

Excluded:
1. Raintight hubs as underground feeders are assumed.

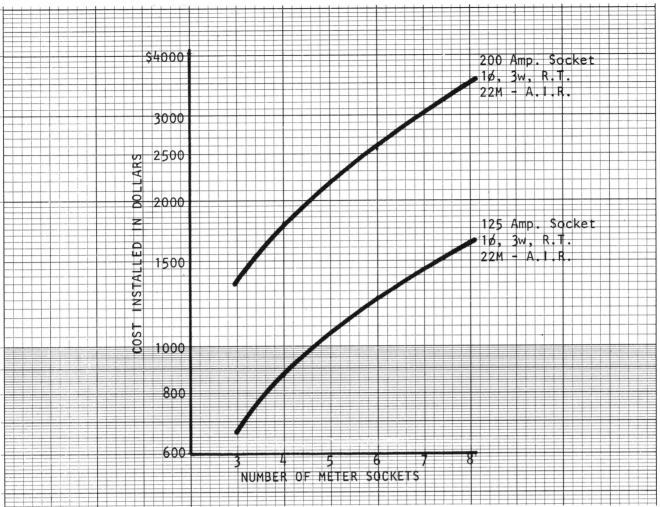

The costs shown for the wall-mounted meter center are based on the published contractors' book prices and are as manufactured by Square D Co. as an outdoor type of meter module.

These meter sockets are single-phase and shown in 100/125-ampere and 200- ampere sizes, the larger generally used for electrically heated apartments. The total installed cost difference between indoor and raintight meter sockets is not significant enough to justify two curves.

Note: While the mains may be single- or three-phase, all services to individual occupancies are single-phase.

Included: (material and labor)

1. Socket type of group and size required.
2. Two pole circuit breakers for each feeder.
3. Fastening devices for fastening to masonry wall.
4. GRC conduit terminals for each meter in the group.
5. Sufficient conductors of length and size to match ampacity and terminal makeup for each feeder.
6. Connection of main buses.

Excluded:

1. Raintight hubs as underground feeders are assumed.

C-7

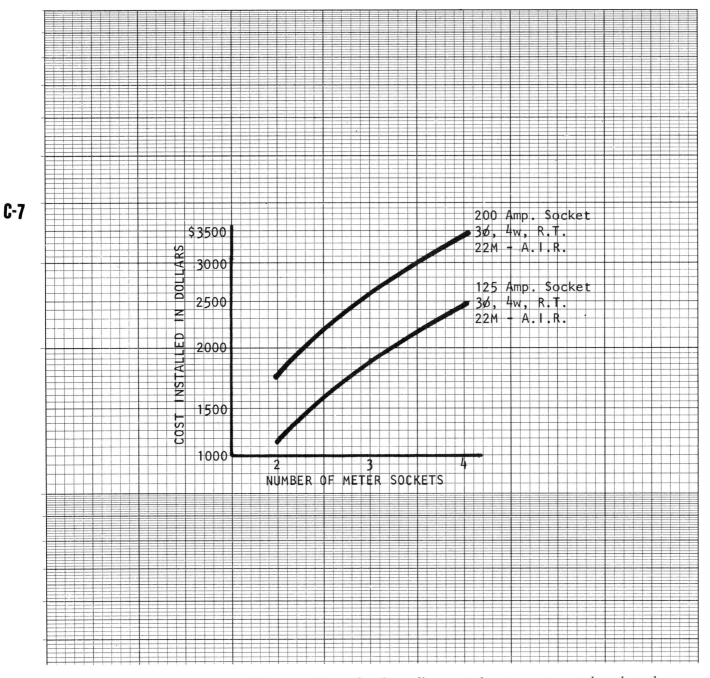

The costs shown for the wall-mounted meter center are based on the published contractors' book prices and are as manufactured by Square D Co. as an outdoor type of meter module.

These meter sockets are three-phase and shown in 125-ampere and 200-ampere sizes. The difference in cost between indoor and outdoor meter sockets is insignificant.

Included: (material and labor)
1. Three-pole circuit breaker for each feeder.
2. Fastening devices for fastening to a masonry wall.
3. GRC conduit terminals for each meter in the group.
4. Sufficient conductors of length and size to match ampacity and terminal makeup for each feeder.

THREE-PHASE EXTERIOR WALL-MOUNTED GROUP METERS

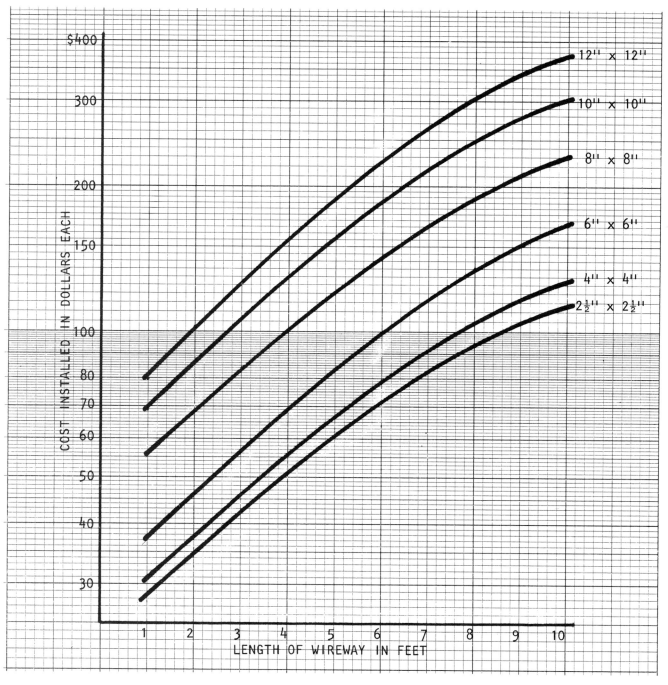

The costs shown for the screw- or hinged-cover wireways consist of the contractors' published book price, fastening devices for mounting on a masonry wall, and the labor required for such installation. The items shown on the graph are manufactured by Hoffman Engineering Co. and contain the necessary couplings and end fittings as required.

C-9

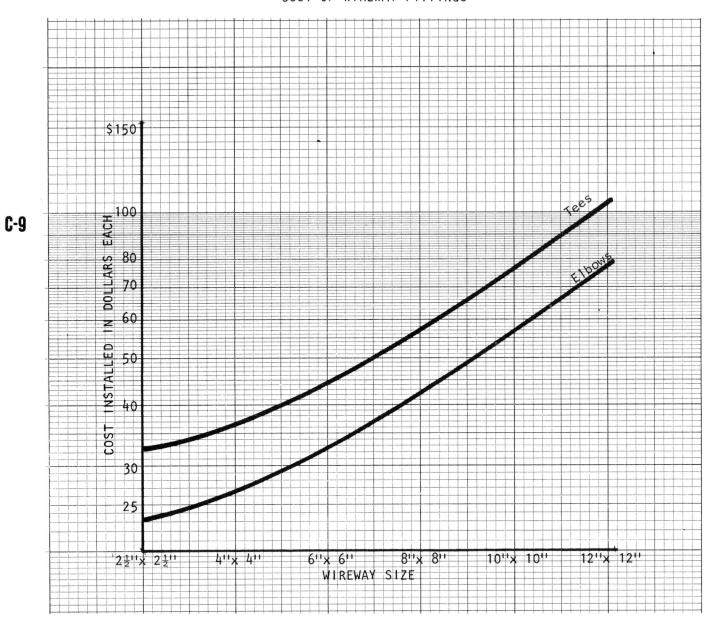

The costs shown for the tees and elbows for the wireway consist of the contractors' published book price, fastening devices for mounting on a masonry wall, required couplings, and the labor for the installation. The items shown are as manufactured by Hoffman Engineering Co.

COST OF RAINTIGHT AND OILTIGHT WIREWAY

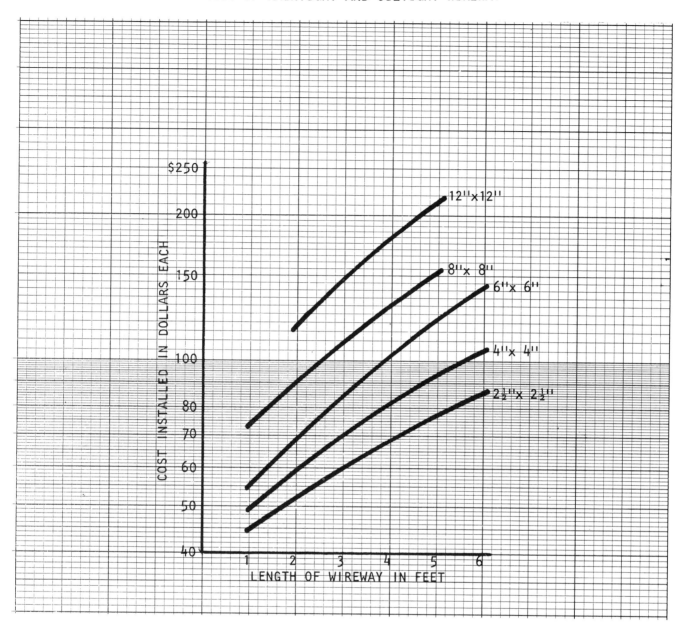

The costs shown for the raintight wireway consist of the published contractors' book price, fastening devices for mounting on a masonry wall, and the labor required for such installation. The items shown are the standard sizes as manufactured by Hoffman Engineering Co. No fittings are available in this series. Covers are gasketed, and there are no conduit knockouts.

Grounding

D

GROUND RODS

Copperclad Ground Rods   D-1
Ground Rod with Lightning-Protection-Type Clamp   D-9
Grounding Plates for Lightning Protection   D-10

GROUND CLAMPS

Ground Clamps for Water Pipe   D-2

GROUNDING CONDUCTORS

Bare Soft-Drawn Copper   D-2
Insulated Copper or Aluminum   E-16
Lightning Protection Types   D-5

EXOTHERMIC CONNECTIONS

Cable-to-Cable Welded Connections   D-3
Cable-to-Steel Connections   D-4

LIGHTNING PROTECTION

Main Conductors—Classes 1 & 2   D-5
Secondary Conductors—Classes 1 & 2   D-5
Air Terminals—Classes 1 & 2   D-6
Connectors—Classes 1 & 2   D-7, D-8, D-9
Ground Plates   D-10
Arresters for Electric Service and TV Antenna   D-10

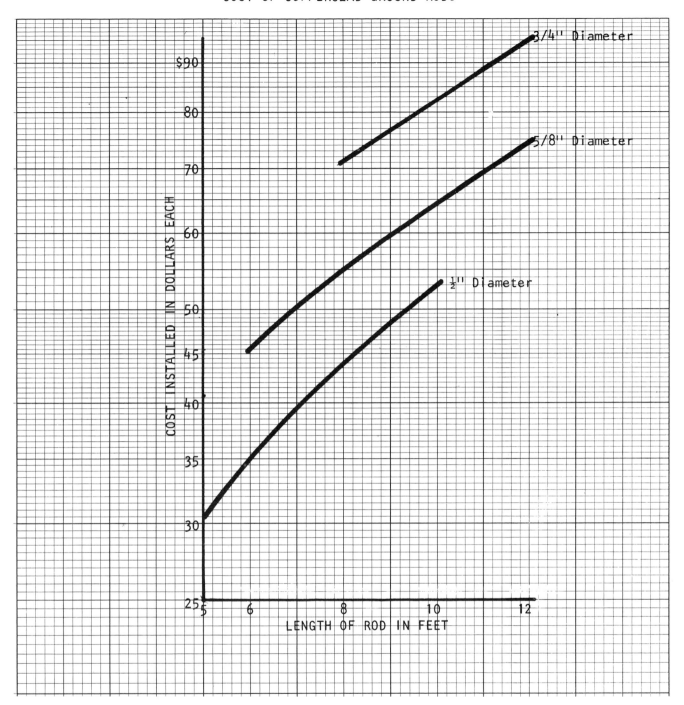

D

D-1

The costs shown for these copperclad ground rods are the contractors'
published book prices for rods as manufactured by Blackburn.

Included: (material and labor)

1. Driving the rod to 6'' below the surface in normal soils.
2. Ground rod clamp for #8 to #1/0 wire.

D-2

COST OF BARE SOFT-DRAWN COPPER

(Y-axis: COST INSTALLED IN DOLLARS PER FOOT — $8.00, 6.00, 5.00, 4.00, 3.00, 2.50, 2.00, 1.50, 1.00, .80)

(X-axis: SIZE OF CONDUCTOR — 8, 6, 4, 2, 1/0, 3/0, 4/0, 300, 400, 500)

The cost shown for the bare copper wire consist of the published contractors' book prices for the sizes shown. The installation labor is based on a single conductor drawn in conduit; the figures will be close enough for most purposes.

COST OF GROUND CLAMPS	
SIZE OF WATER PIPE	COST INSTALLED
1/2″, 3/4″, and 1″	$11.50
1 1/4″, 1 1/2″, and 2″	20.00
2 1/2″, 3″, 3 1/2″, and 4″	36.75

The costs shown for ground clamps are based upon T&B bronze clamps and the labor to attach to the pipe size shown and connect the ground wire,

Form of Connection	Installed Cost
 **SS** Splice/Horizontal	$44.00
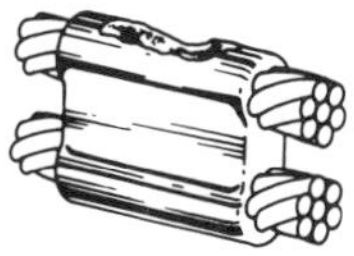 **PT** Parallel thru/Horizontal — Over and under	45.25
 **TA** Tee/Horizontal	45.00
 **XA** X/Horizontal — Same plane	45.50
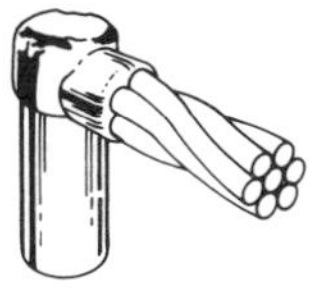 **GR** Ell/Horizontal cable to vertical down rod — Right angle	36.50
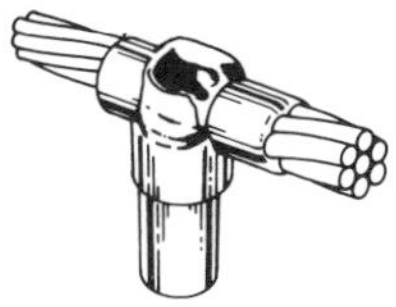 **GT** Tee/Horizontal cable, Vertical rod	37.00

D

D-3

The cost of the welded connections shown includes the published contractors' book price for the weld metal required, one-fiftieth of the cost of the mold (the mold lasts for about 50 shots), and the labor of preparation. The prices shown are the average for copper wire sizes from #6 through #4/0.

These Cadweld grounding connections are as manufactured by Erico Products Inc. of Cleveland, Ohio.

Form of Connection	Installed Cost
① **HA** Tap/Horizontal — To steel, on surface ① **HB** Tap/Horizontal — To cast iron	$44.50
① **VF** Tap/Vertical — To steel, cable up and on surface ① **VK** Tap/Vertical — To cast iron, cable up	46.00
VS Tap/Vertical — Down at 45° to steel	44.75
① **VN** Tap/Horizontal — To steel, on surface. Specify right or left hand. Right hand shown. ① **VR** Tap/Horizontal — To cast iron. Specify right or left hand. Right hand shown.	44.75

D-4

The cost of the welded connections shown includes the published contractors' book price for the weld metal required, one-fiftieth of the cost of the mold (the mold lasts for about 50 shots), and the labor of preparation. The prices shown are the average for copper wire sizes from #6 through #4/0.

These Cadweld grounding connections are as manufactured by Erico Products Inc. of Cleveland, Ohio.

D

*The installation of a lightning protection system which will provide
the caliber of protection an owner rightfully expects involves knowledge
not generally possessed by the usual electrical contractor. It is desirable,
therefore, that both the designer and the contractor become acquainted
with some of the system's installation peculiarities. For further
information refer to:*

Lightning Protection Institute
Harvard, Illinois 60033
815-943-7211

    Programs:
      1. The School of Lightning Protection Technology
      2. Testing and certification
      3. Standards
      4. Consumer education
      5. Industry ethics
      6. Certified system

National Fire Protection Association
Batterymarch Park
Quincy, Mass. 02269

    Lightning Protection Code, 1983, NFPA-78

*The costs shown include the published contractors' book price and
labor for installation under normal conditions. The equipment shown
is as manufactured by Thompson Lightning Protection Inc. of St. Paul,
Minn.*

D-5

CLASS I

An ordinary building is one of common or conventional design and construction used for ordinary purposes, whether commercial, farm, institutional, industrial or residential. A Class I ordinary building is one which is less than 75 feet in height.

CLASS II

A Class II ordinary building is one more than 75 feet in height. The distinction in terms of lightning protection is that air terminals, conductors, and ground rods of Class II structures are of larger dimensions and higher conductance than minimum allowances for Class I buildings.

Description	Cost installed in dollars per foot		Description	Cost installed in dollars per foot	
	Copper	Aluminum		Copper	Aluminum
MAIN CONDUCTOR Cu: 57,400 CM Min. Al: 98,600 CM Min. Exposed on wood	$2.45	$1.97	MAIN CONDUCTOR Cu: 115,000 CM Al: 192,000 CM Exposed on wood	$3.15	$2.25
MAIN CONDUCTOR Exposed on masonry	$3.30	$2.84	MAIN CONDUCTOR Exposed on masonry	$4.00	$3.12
MAIN CONDUCTOR Adhesive cable holder	$3.10	$2.47	MAIN CONDUCTOR Adhesive holder	$4.68	$3.62
MAIN CONDUCTOR In free air	$1.63	$1.17	MAIN CONDUCTOR In free air	$2.28	$1.45
SECONDARY CONDUCTOR Cu: 26,240 CM Min. Al: 41,100 CM Min. With 6' cable, clamp and terminal	$23.50 each	$19.75 each	SECONDARY CONDUCTOR Cu: 26,240 CM Al: 41,100 CM With 6' cable, clamp & terminal	$28.20 each	$15.50 each

LIGHTNING PROTECTION—CABLE

COST OF AIR TERMINALS - LIGHTNING PROTECTION
PROTECTION OF ORDINARY BUILDINGS
ANSI/NFPA 78, 1983

CLASS I

An ordinary building is one of common or conventional design and construction used for ordinary purposes, whether commercial, farm, institutional, industrial or residential. A Class I ordinary building is one which is less than 75 feet in height.

CLASS II

A Class II ordinary building is one more than 75 feet in height. The distinction in terms of lightning protection is that air terminals, conductors, and ground rods of Class II structures are of larger dimensions and higher conductance than minimum allowances for Class I buildings.

Description	Cost installed in dollars each		Description	Cost installed in dollars each	
	Copper	Aluminum		Copper	Aluminum
AIR TERMINAL Surface mtd.	$45.50	$41.25	AIR TERMINAL Surface mtd.	$66.50	$59.00
AIR TERMINAL Concealed base	$63.25	$60.75	AIR TERMINAL Concealed base	$78.75	$72.00
AIR TERMINAL Adhesive base	$46.25	$38.50	AIR TERMINAL Adhesive base	$59.25	$50.00
AIR TERMINAL Parapet base	$45.00	$39.25	AIR TERMINAL Parapet base	$53.25	$51.00
AIR TERMINAL Chimney type	$55.50	$46.00	AIR TERMINAL Chimney type	$64.50	$47.75

D-7

CLASS I			CLASS II		
An ordinary building is one of common or conventional design and construction used for ordinary purposes, whether commercial, farm, institutional, industrial or residential. A Class I ordinary building is one which is less than 75 feet in height.			A Class II ordinary building is one more than 75 feet in height. The distinction in terms of lightning protection is that air terminals, conductors, and ground rods of Class II structures are of larger dimensions and higher conductance than minimum allowances for Class I buildings.		
Description	Cost installed in dollars each		Description	Cost installed in dollars each	
	Copper	Aluminum		Copper	Aluminum
BUS TERMINAL	$22.25	$21.75	BUS TERMINAL	$21.75	$20.75
METAL BONDING PLATE	$19.00	$18.25	METAL BONDING PL.	$30.25	$29.00
SILL COCK	$23.25	$22.25	BONDING PLATE TO STRUCTURAL STEEL	$31.00	$29.25
THRU ROOF OR WALL	$58.50	$58.00	THRU ROOF OR WALL	$67.50	$61.25
EAVE TROUGH	$8.25	$7.75	N.A.		

COST OF CONNECTORS - LIGHTNING PROTECTION
PROTECTION OF ORDINARY BUILDINGS
ANSI/NFPA 78, 1983

CLASS I	CLASS II
An ordinary building is one of common or conventional design and construction used for ordinary purposes, whether commercial, farm, institutional, industrial or residential. A Class I ordinary building is one which is less than 75 feet in height.	A Class II ordinary building is one more than 75 feet in height. The distinction in terms of lightning protection is that air terminals, conductors, and ground rods of Class II structures are of larger dimensions and higher conductance than minimum allowances for Class I buildings.

Description	Cost installed in dollars each		Description	Cost installed in dollars each	
	Copper	Aluminum		Copper	Aluminum
CABLE TO I BEAM	$13.25	$11.50	CABLE TO I·BEAM	$13.25	$11.50
PIPE STRAP	$15.25	$13.50	PIPE STRAP	$25.00	$19.00
COPPER-TO-ALUMINUM CABLE	$10.00	$10.00	COPPER-TO-ALUMINUM CABLE	$10.00	$10.00
DISCONNECTOR FOR GROUNDING	$12.00	- - -	DISCONNECTOR FOR GROUNDING	$13.00	- - -

D-9

CLASS I

An ordinary building is one of common or conventional design and construction used for ordinary purposes, whether commercial, farm, institutional, industrial or residential. A Class I ordinary building is one which is less than 75 feet in height.

CLASS II

A Class II ordinary building is one more than 75 feet in height. The distinction in terms of lightning protection is that air terminals, conductors, and ground rods of Class II structures are of larger dimensions and higher conductance than minimum allowances for Class I buildings.

Description	Cost installed in dollars each		Description	Cost installed in dollars each	
	Copper	Aluminum		Copper	Aluminum
STRAIGHT SPLICER	$4.25	$3.75	STRAIGHT SPLICER	$12.25	$9.75
PARALLEL SPLICER	$4.25	$3.75	PARALLEL SPLICER	$10.00	$9.00
TEE SPLICER	$4.25	$3.75	TEE SPLICER	$13.00	$10.00
GROUND ROD & CLAMP ½" x 10'	$50.25	N.A.	GROUND ROD & CLAMP 5/8" x 10'	$52.00	N.A.
WATER-PIPE CLAMP	$21.75	$20.50	WATER-PIPE CLAMP	$33.50	$31.50

Description	Cost installed in dollars each
GROUND PLATES For areas where rods are impossible to drive due to rock conditions	$104.00
ELECTRIC SERVICE Single phase, 3 wire, 120/240 volt	65.50
Three phase, 4 wire, 120/208 volt	103.00
TELEVISION ANTENNA	45.00

FLAT PLATES

Arrester

TV ANTENNA PROTECTION

1: APPROVED CLAMP WITH 1½" CONTINUOUS CONTACT WITH MAST AND FULL SIZE CABLE.
2: LEAD IN ARRESTER.

D

D-10

Feeders

CONDUITS

Electric Metallic Tubing—EMT
Galvanized Rigid Steel—GRC
Aluminum Rigid Conduit—ARC
Polyvinyl Chloride—Rigid Conduit—PVC
Flexible Metal—FLEX.
Liquidtight Flexible Metal
PVC-Coated Galavanized Rigid Conduit
Intermediate Metal Conduit—IMC   E-1

CONDUIT FITTINGS

Terminals   E-2
Fittings: Seal-Offs, LBs, LLs, Entrance Caps   E-3
Couplings—EMT, ARC, GRC, PVC, Erickson   E-4
Nipples—Straight, Offset, and Chase   E-5
Field Bends   E-6
Feeder Conduit Factory Elbows   E-7

CONDUCTORS

Copper
THW Insulation   E-8
XHHW Insulation   E-9
THWN/THHN Insulation   E-10
USE, XLPE—Direct Burial   E-11
Service Entrance Cable—SE   E-12
Bare Copper Wire   D-2
Nonmetallic Sheathed Cable—NM   F-11
Armored Cable—(BX)   F-14
Grounding Conductor—Insulated   E-16
Aluminum
XHHW & THW   E-14
USE/XLPE—Direct Burial   E-15
Self-Supporting Service Drop Cable   E-13
Service Entrance Cable—SE   E-12
Grounding Conductor—Insulated   E-16

CONDUCTOR TERMINALS AND TAPS

Compression Terminals for Copper or Aluminum
and Split Bolt Tap Connectors   E-17

E

FEEDER BUSWAY

Copper and Aluminum Feeder Busway   E-18
Copper and Aluminum Feeder Busway Transformer Taps   E-19
Copper and Aluminum Feeder Busway Elbows   E-20
Copper and Aluminum Feeder Busway Tees   E-21
Copper and Aluminum Feeder Busway Terminals   E-22

PLUG-IN BUSWAY

Copper and Aluminum Plug-In Busway   E-23
Copper and Aluminum Busway Elbows   E-24
Copper and Aluminum Busway Tees   E-25
Copper and Aluminum Plug-in Busway Cable Tap Box   E-26
Copper and Aluminum Plug-in Busway Switches   E-27

DUCT BANKS

Direct Burial—Earth Cover—Plastic Duct   E-28
Concrete-Encased—Plastic Duct   E-29
Reinforced-Concrete-Encased—Plastic Duct   E-30

LADDER TRAYS—4" DEEP

Steel Tray   E-31
Horizontal Elbow—45° and 90°   E-31
Horizontal Tee   E-32
Vertical Elbow   E-32
Straight Reducer   E-32

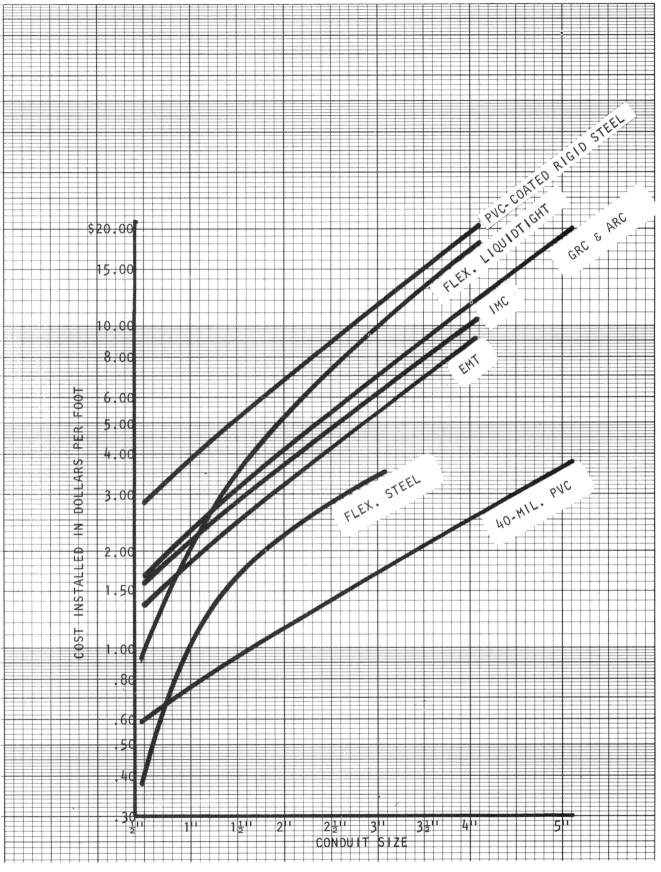

The costs shown for these conduits consist of the published contractors'
book price and the necessary labor for installation exposed on wood at
a height not exceeding 12 feet. No ground wire is included in PVC conduit.
If necessary, see page E-16.

E-2

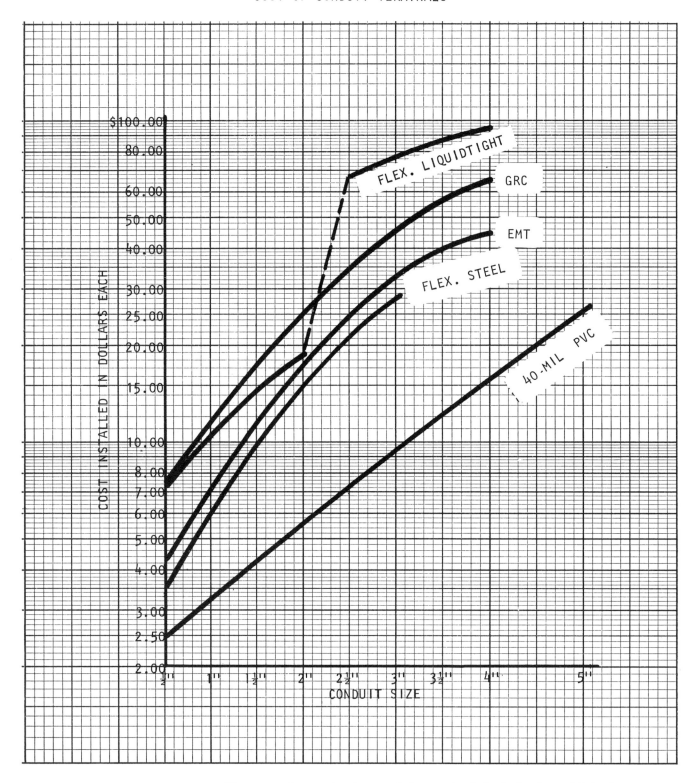

The costs shown for the conduit terminals include the published
contractors' book price for the terminals shown. They also include the
labor to cut and thread the conduit to the required length and installation
of the terminal to the box.

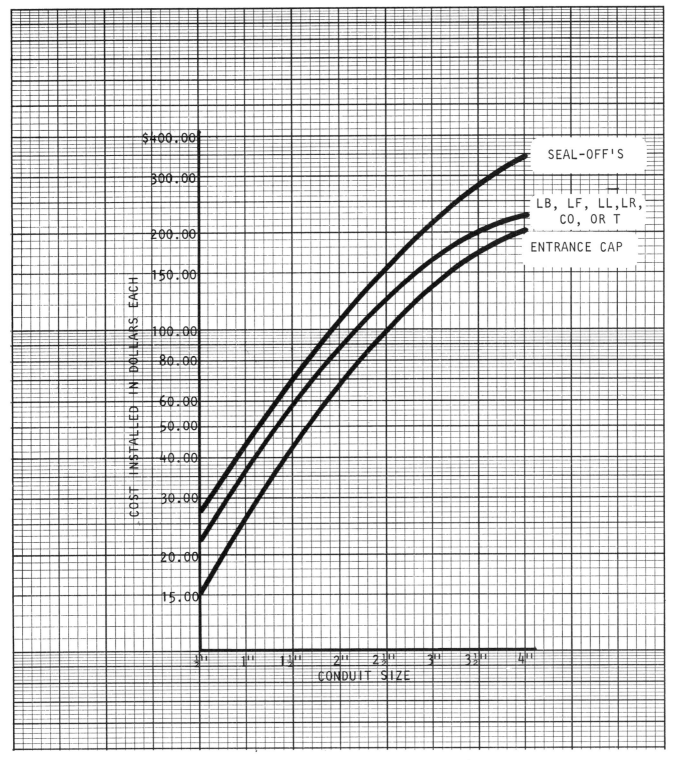

The cost shown for the conduit fittings includes the published contractors'
book price for the various fittings shown of the gasketed type to suit the
kind of conduit required. It also includes the labor of preparing the
end of the conduit and installing the fitting. Seal-offs are considered to
be used with GRC in hazardous applications, and include the installation
of the sealing compound as required.

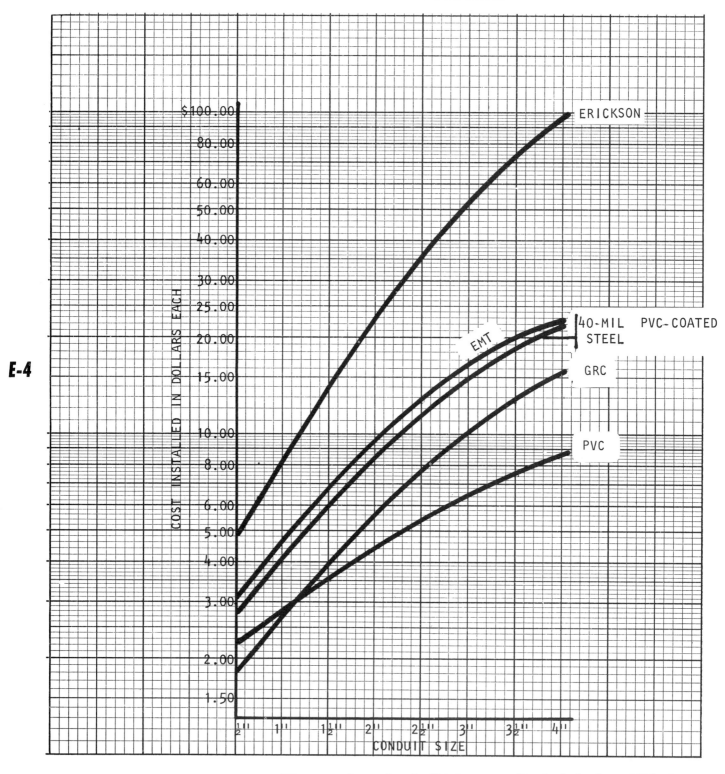

The costs shown for the Erickson coupling include the published contractors' book price of the sizes included and sufficient labor for cutting and threading GRC conduit to accept the coupling.

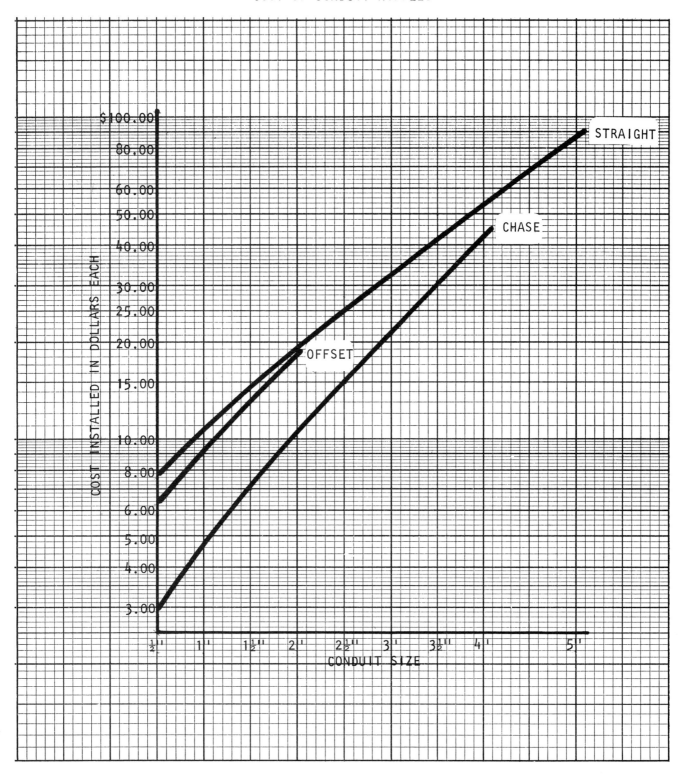

The costs shown on the graph for conduit nipples consist of the cost of
a 4-inch-long GRC nipple with two locknuts and a metal-insulated bushing
at each end of the nipple. The labor includes installation of the nipple
but not cutting the holes in the steel panel. See page L-5 for cutting
holes in a steel panel.

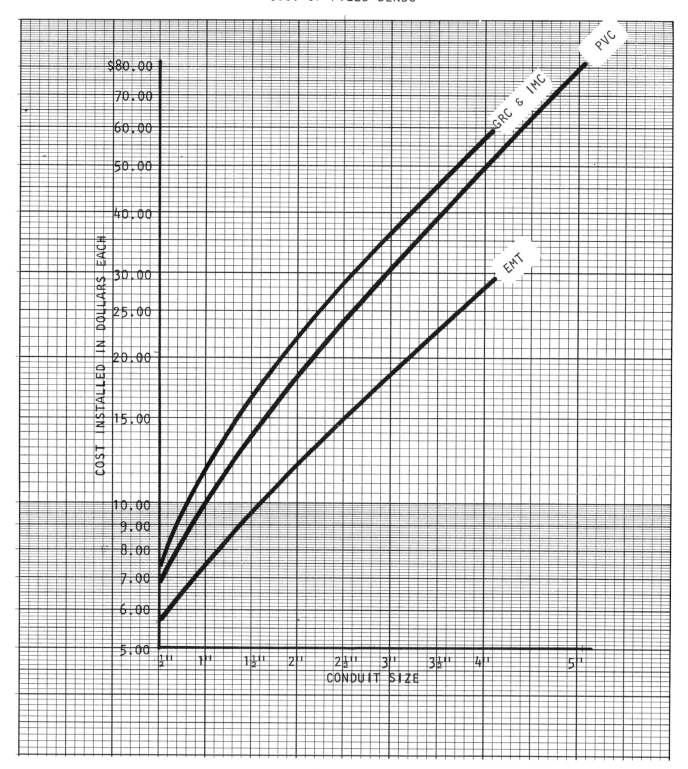

The costs shown for the field bends have no material associated with them; however, they do include the labor for making bends with a hydraulic bender or a heat type of bender for the PVC.

E-6

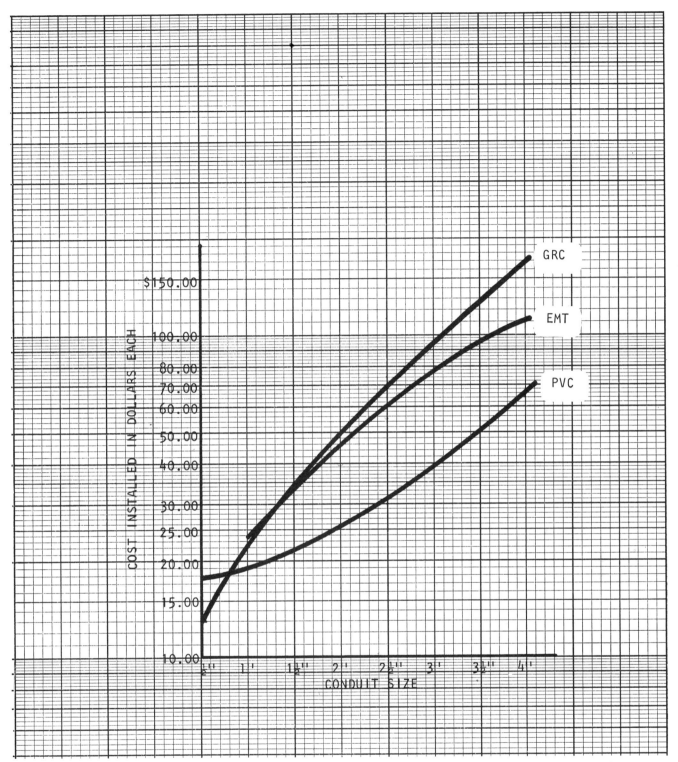

The costs shown on this graph for conduit elbows include the published contractors' book price for factory-made elbows. They also include one coupling and the labor for cutting and threading one end of the conduit to fit the elbow as required.

E

E-7

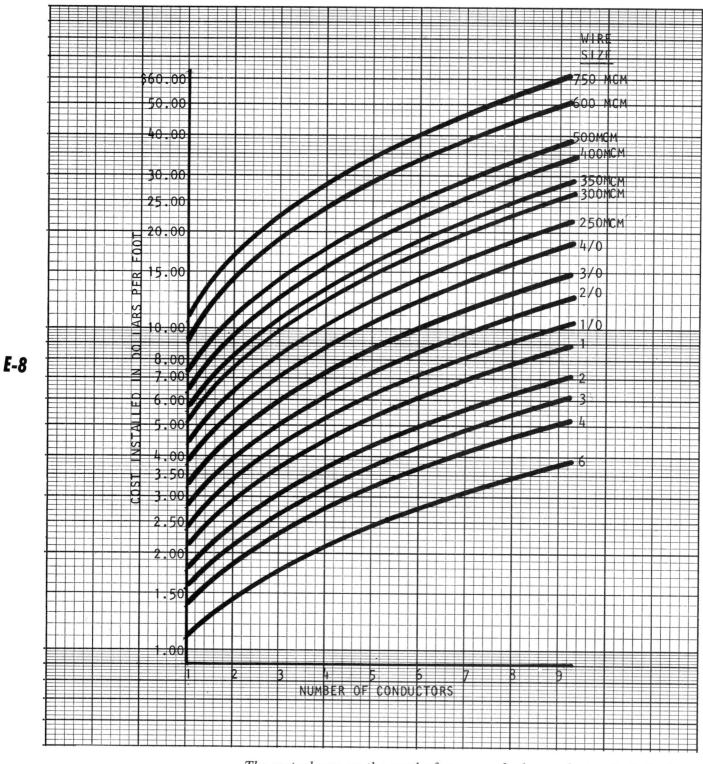

E-8

The costs shown on the graphs for copper feeder conductors include the published contractors' book price for the copper wire and insulation indicated on the graphs. The installation labor is predicated on an average pull of 100 feet in conduit.

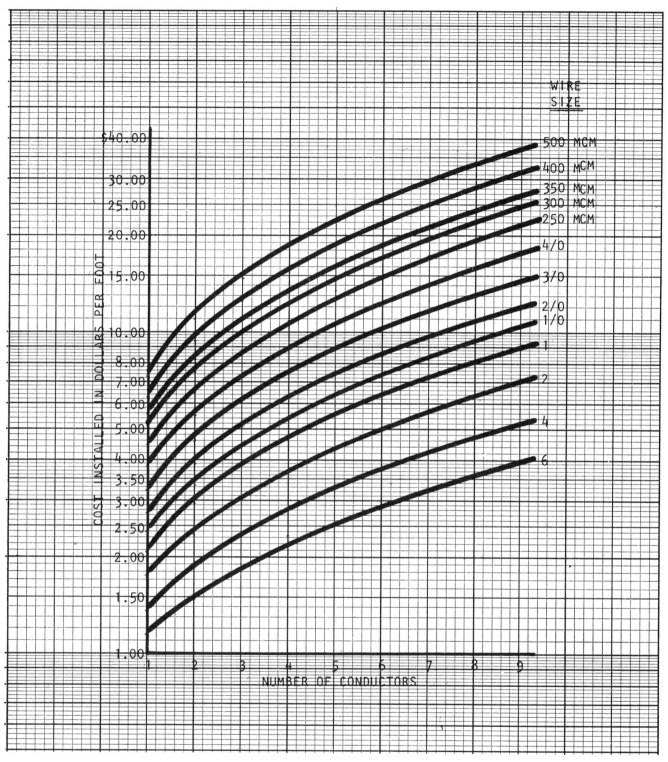

The costs shown on the graphs for copper feeder conductors include the
published contractors' book price for the copper wire and insulation
indicated on the graphs. The installation labor is predicated on an average
pull of 100 feet in conduit.

E-10

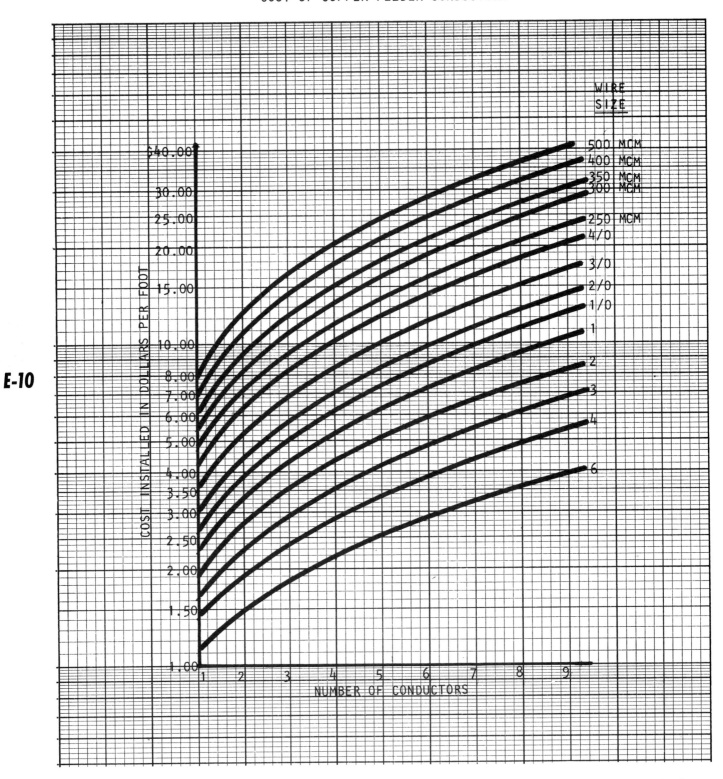

The costs shown on the graphs for copper feeder conductors include the the published contractors' book price for the copper wire and insulation indicated on the graphs. The installation labor is predicated on an average pull of 100 feet in conduit.

THWN/THHN - COPPER - FEEDER CONDUCTORS

COST OF COPPER FEEDER CONDUCTORS

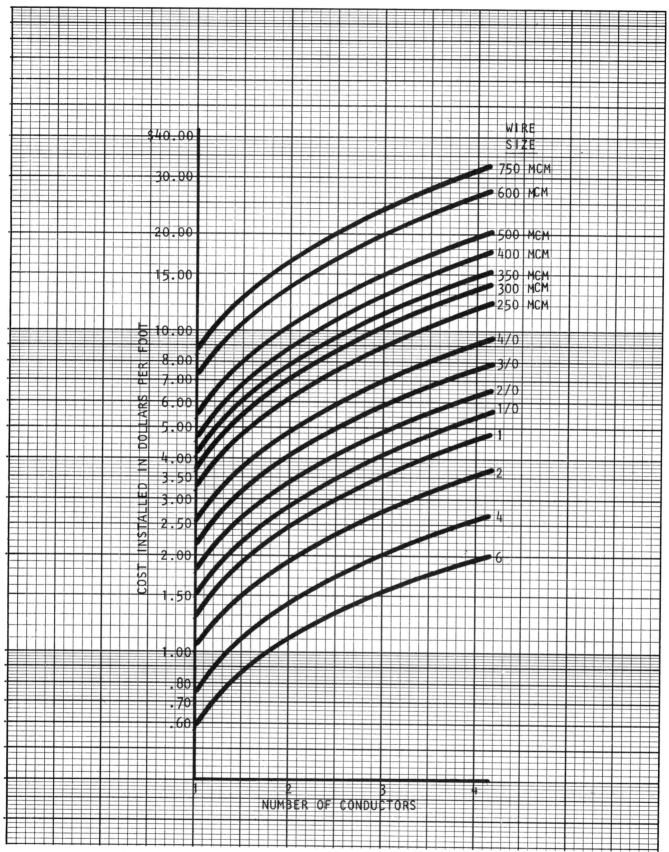

The costs shown on the graph for copper USE/XLPE conductors include
the published contractors' book price for the copper wire and the
insulation indicated on the graph. The installation labor is predicated
upon laying conductors in an open trench. Trenching costs are not
included; for these see page L-2.

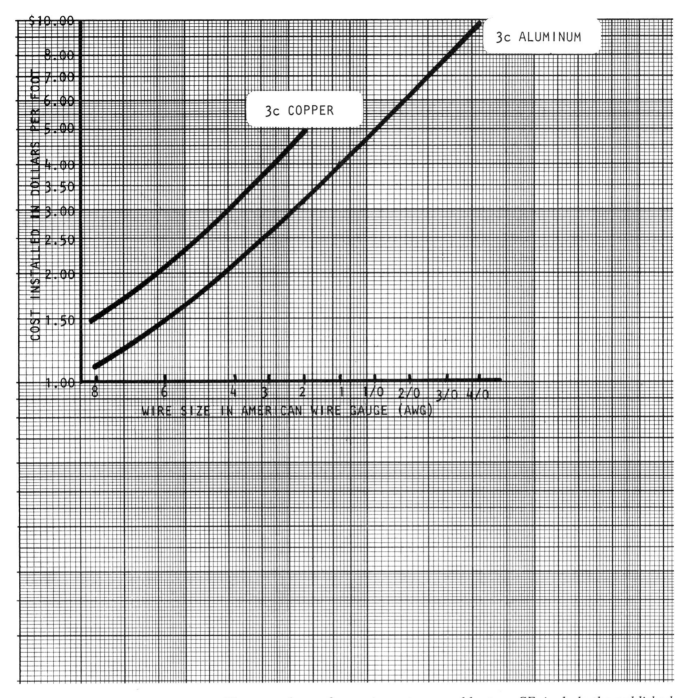

E-12

The costs shown for service entrance cable, type SE, include the published contractors' book price for the cable shown and straps for supporting it on 4-foot centers. Also included is the labor for installation on exposed wood surfaces.

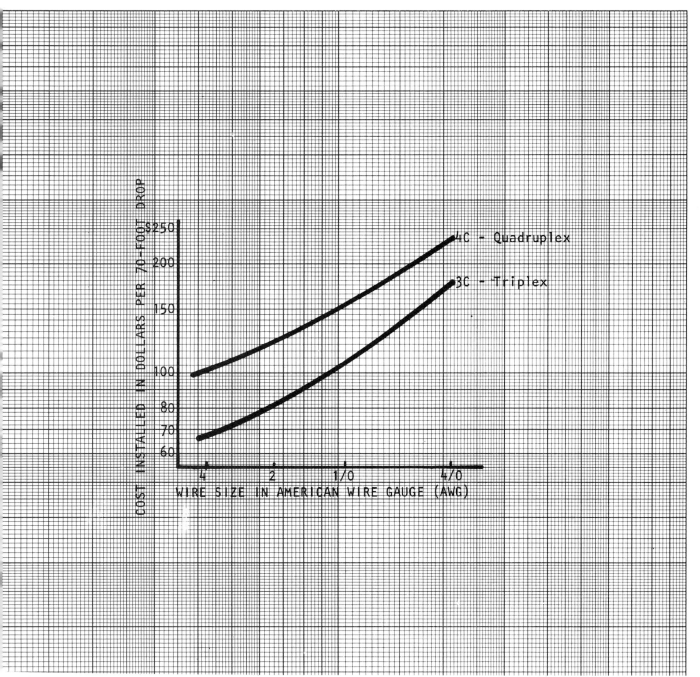

E

E-13

The costs shown on the graph for self-supporting service drop cable
include the published contractors' book price for the aluminum wire
with a steel messenger for support and the labor for installing 70 feet
of wire and the anchors required on the pole and building at each end.
The 70-foot length is used as an average.

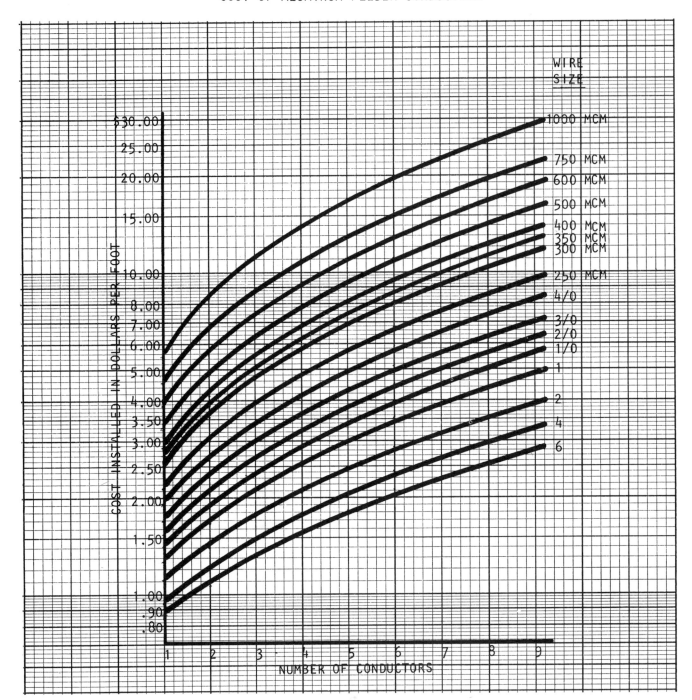

The costs shown on the graphs for aluminum feeder conductors include the published contractors' book price for the aluminum wire and insulation as indicated on the graphs. The installation labor is predicated upon an average pull of 100 feet in conduit.

E-14

COST OF ALUMINUM FEEDER CONDUCTORS

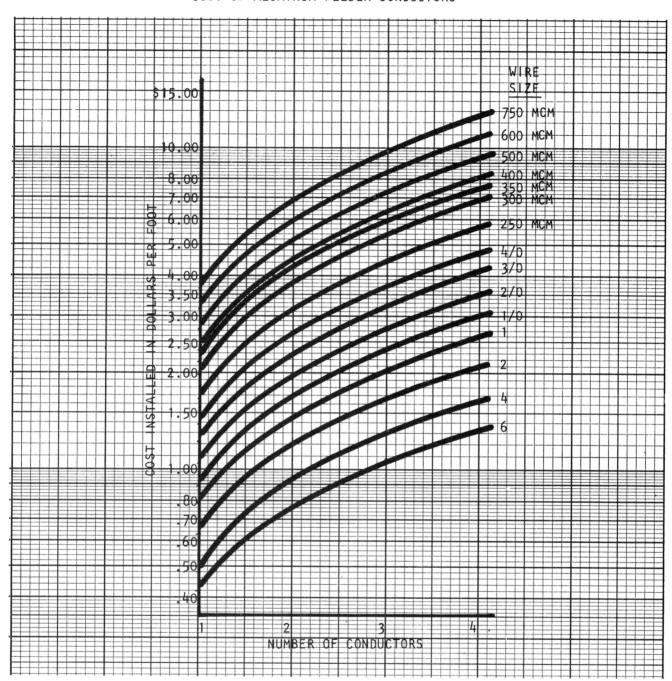

The costs shown on the graphs for aluminum feeder conductors include the published contractors' book price for the aluminum wire and insulation as indicated on the graphs. The installation labor is predicated upon laying conductors in an open trench. The cost of digging the trench is not included. See page L-2.

E-16

The costs shown on the graph for insulated grounding conductors include the published contractors' book price for the type of wire indicated on the graph. The installation labor is predicated upon pulling a single conductor along with four-phase conductors of the size normally used for the size of grounding conductor required.

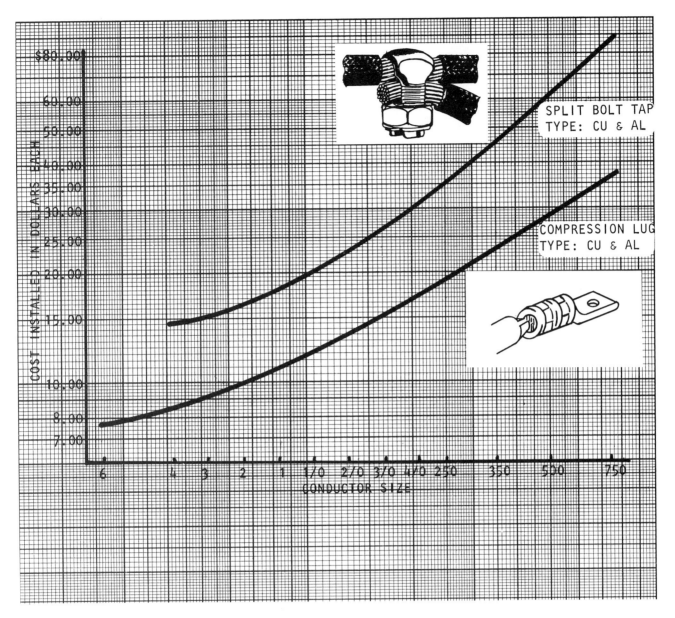

The costs shown for the aluminum compression type of terminal include the published contractors' book price for the Thomas & Betts wrought-aluminum tin-plated terminal prefilled with oxide-inhibiting compound. The labor includes cutting and stripping the cable and installing the terminal with the proper manual-type compression tools.

The costs shown for the copper split bolt tap connector include the published contractors' book price for the Burndy split bolt solderless connector and electrical tape. Labor is provided for cable preparation and taping.

E-18

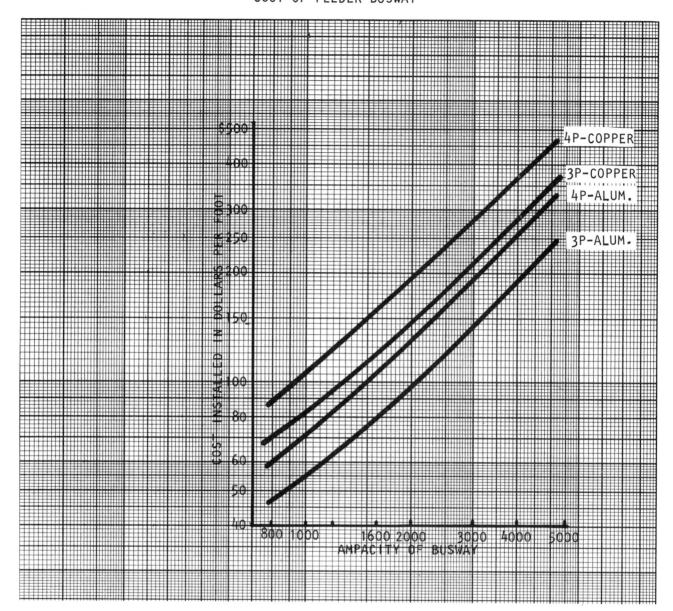

The costs shown on this graph for copper and aluminum feeder busway consist of the contractors' published book prices and two 36-inch hangers for every 10 feet of busway. Labor is included to mount the hangers and install the busway. Four-wire busway has full neutral.

COST OF FEEDER BUSWAY TRANSFORMER TAPS

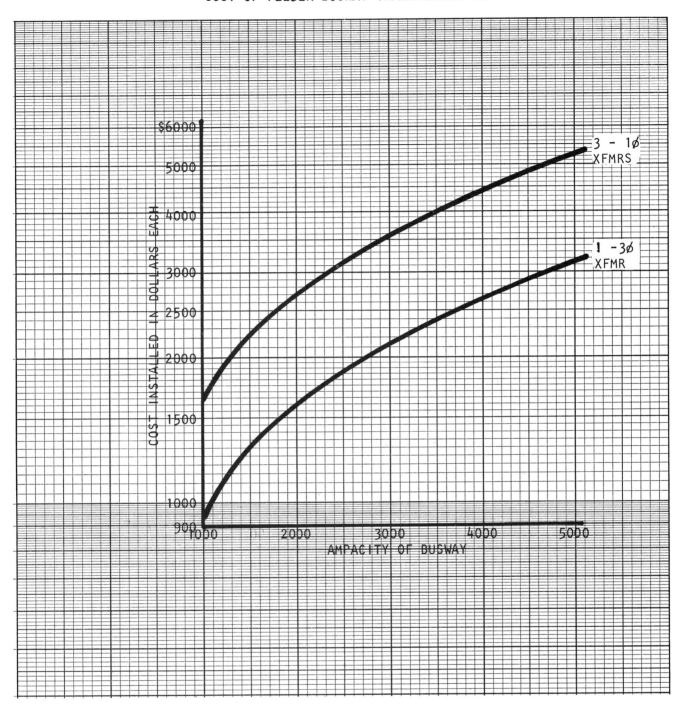

The costs shown for the feeder busway transformer taps consist of the contractors' published book price for the taps indicated. The cost of the busway length must be added to these. The three-phase tap from a single transformer has four busway connections—three hot legs and a neutral. For three single-phase transformers there are six busway connections—two for each transformer. Included in these costs shown are 5-foot drops of copper, in the ampacity of the busway, and copper terminals at each end of the drop, with labor to make up the terminations and install the busway tap and wire.

COST OF FEEDER BUSWAY ELBOWS

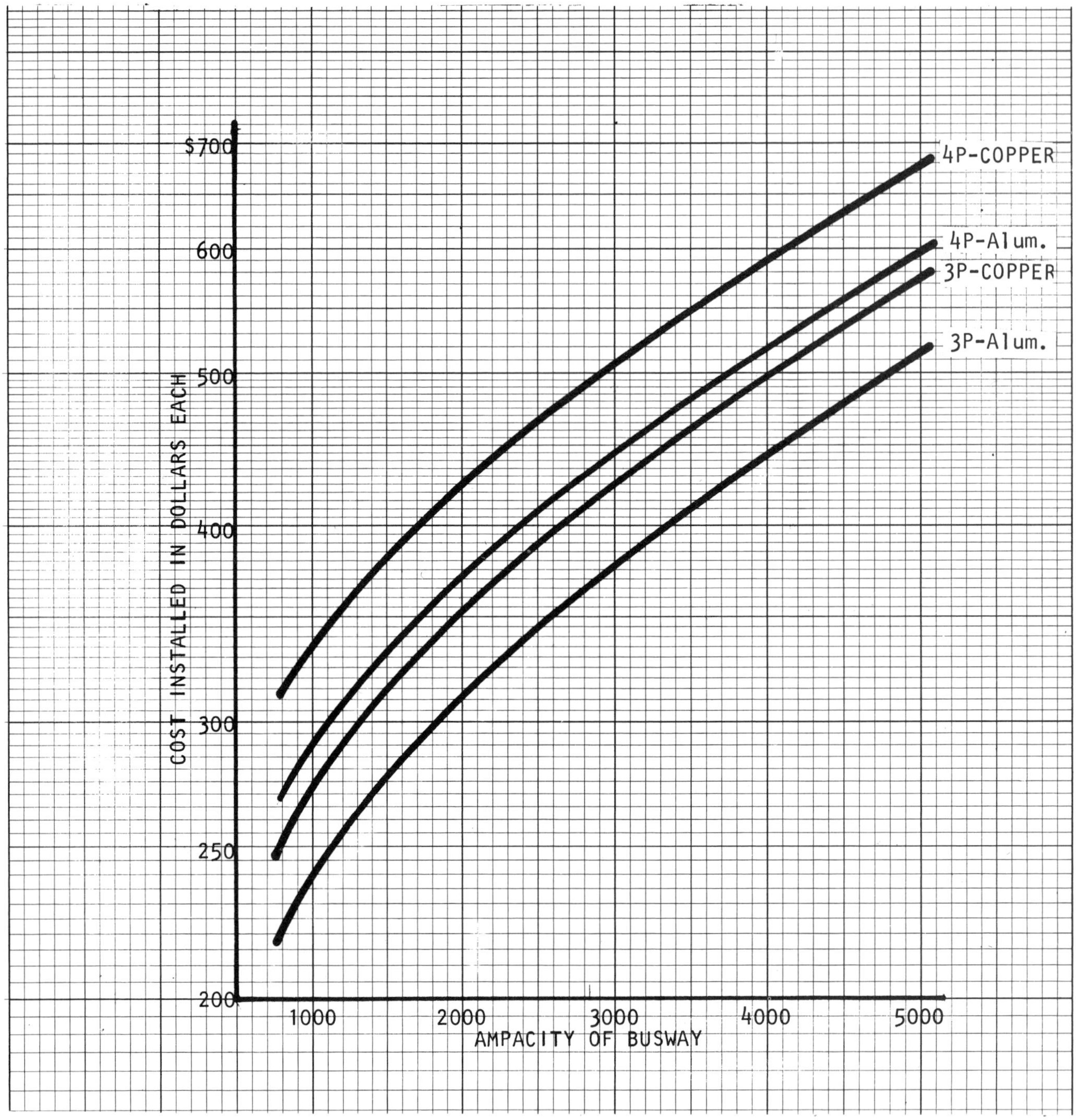
$700
600
500
400
300
250
200
COST INSTALLED IN DOLLARS EACH
1000
2000
3000
4000
5000
AMPACITY OF BUSWAY
4P-COPPER
4P-Alum.
3P-COPPER
3P-Alum.

The costs shown on this graph for feeder busway elbows consist of the contractors' published book price for a "labor only" charge by the manufacturer. The length of the busway duct through the fitting must be added to these elbow charges. The cost of the labor required by the contractor for assembling and hanging the duct is included in the prices shown on the graph.

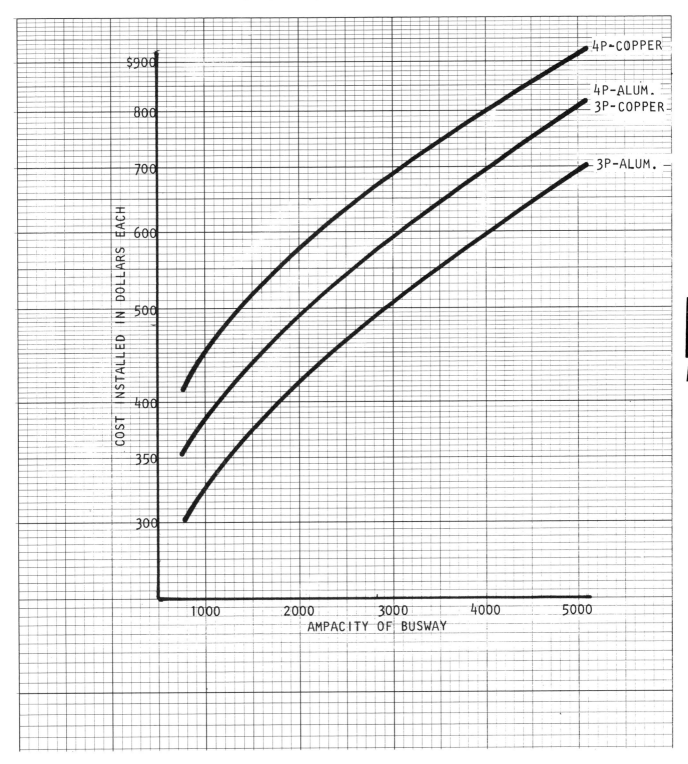

The costs shown on this graph for feeder busway tees consist of
the contractors' published book price for a ''labor only'' charge by the
manufacturer. The length of the busway duct through the fitting must
be added to these tee charges. The cost of the labor required by the
contractor for assembling and hanging the duct is included in the prices
shown on the graph.

The costs shown on this graph for the feeder busway terminal consist of the contractors' published book price for a "labor only" charge by the manufacturer. The footage of the busway through the fitting must be added to these terminal charges. The cost of the labor required by the contractor for assembling and hanging the duct is included in the prices shown on the graph.

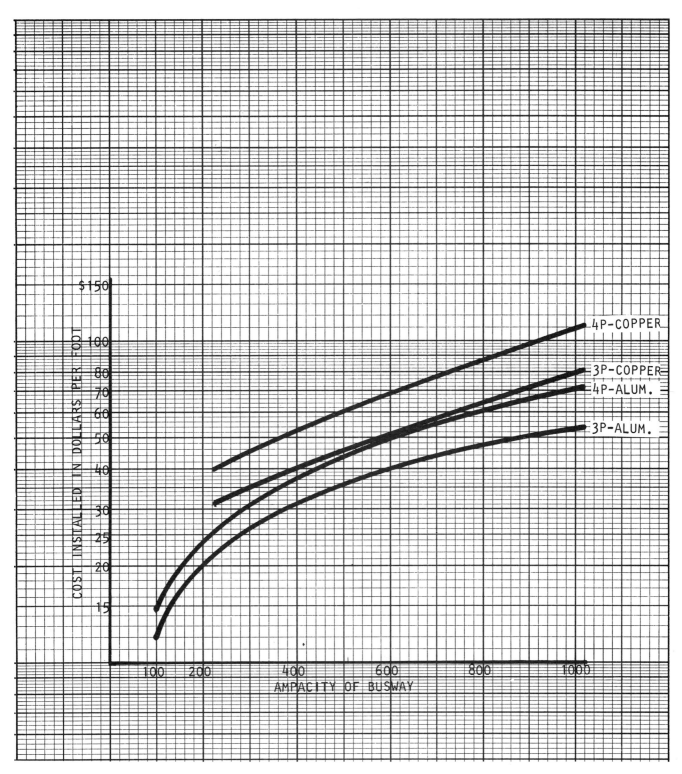

*The costs shown for the plug-in busway consist of the contractors'
published book price and include one 36-inch hanger rod every 5 feet
with a fitting for fastening to a steel bar joist. The duct is full neutral
of the types shown, and labor for installing the busduct is included.*

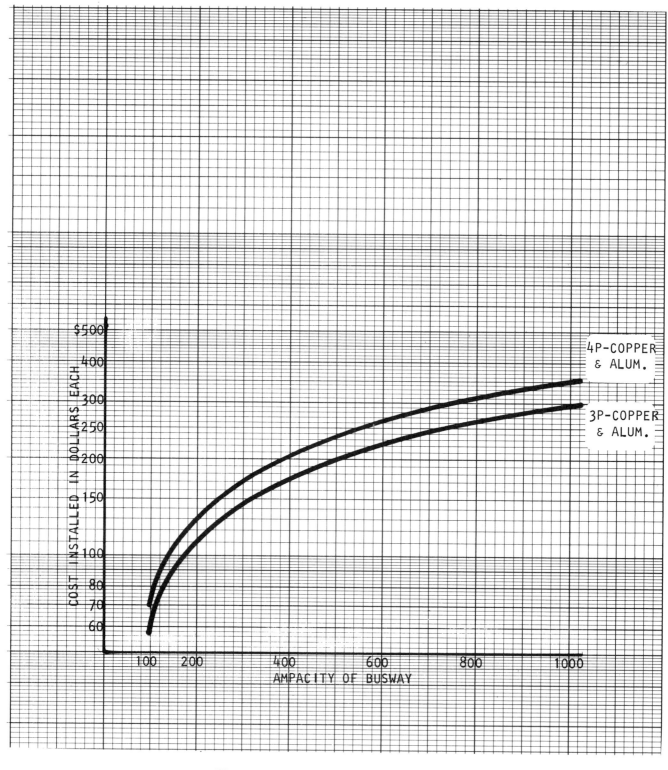

E-24

*The costs shown for the busway elbows consist of the contractors'
published book price and are the manufacturer's labor charges for
making the elbows. The busway cost must be included with the busway
footage by measuring through these elbows. The costs shown here
include the contractor's charge for installing the busduct elbows as well
as the contractor's cost for the elbows.*

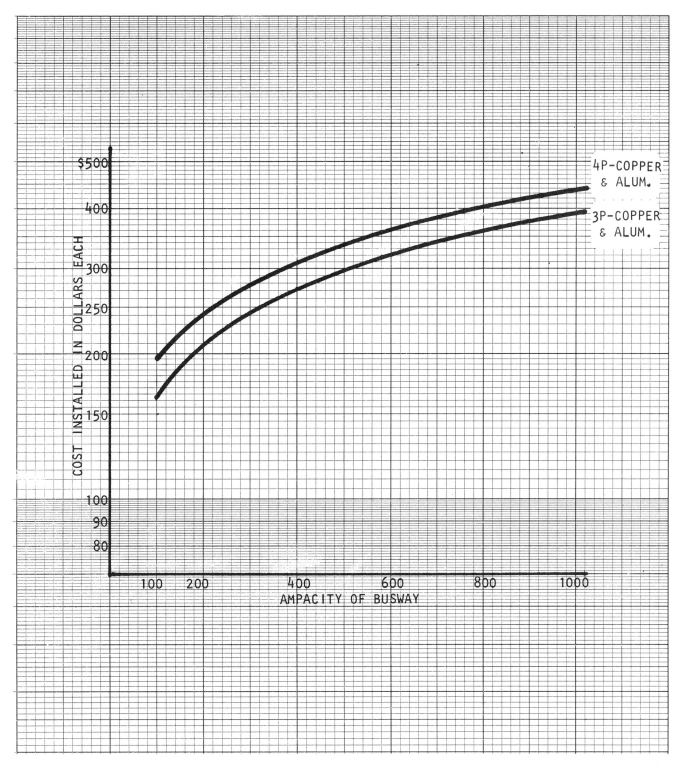

E

E-25

The costs shown for the busway tees consist of the contractors' published book price and are the manufacturer's labor charges for making the fittings. The busway costs must be included with the busway footage by measuring through the fittings. The costs shown here include the contractor's charge for installing the busduct fitting as well as the contractor's cost for the fitting.

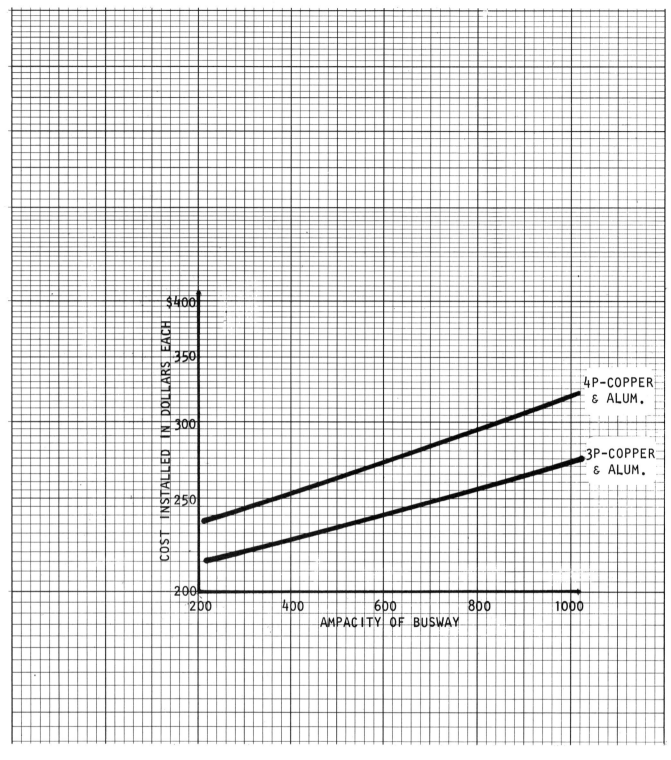

E-26

*The costs shown for the busway cable tap box consist of the contractors'
published book price and are the manufacturer's labor charges for making
the fitting. The busway cost must be included with the busway footage
by measuring through these fittings. The cost of the labor by the
contractor for assembling and hanging the busway fitting is included
in the prices shown on the graph.*

COST OF PLUG-IN BUSWAY SWITCHES

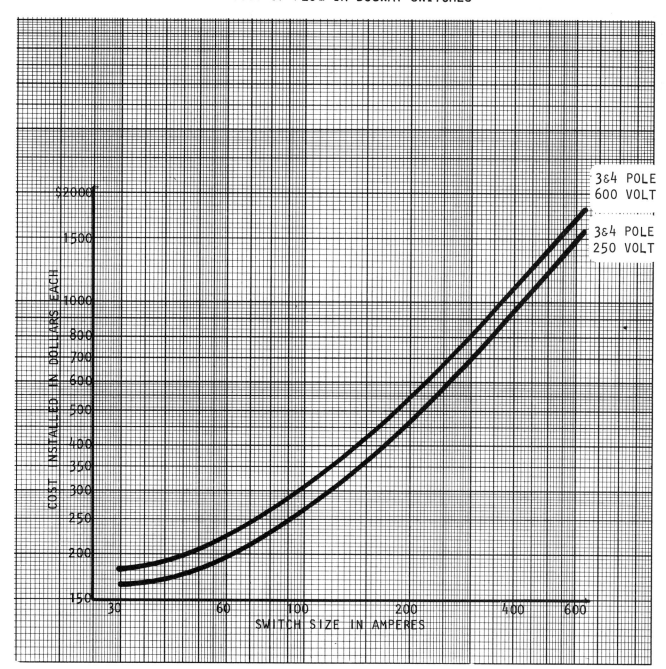

The costs shown for the plug-in busway switches in the 250- and 600-volt classes consist of the published contractors' book prices with the necessary one-time fuses and installation on the busway. The costs shown do not include conduit terminals or the conductor length required within the switch.

E-28

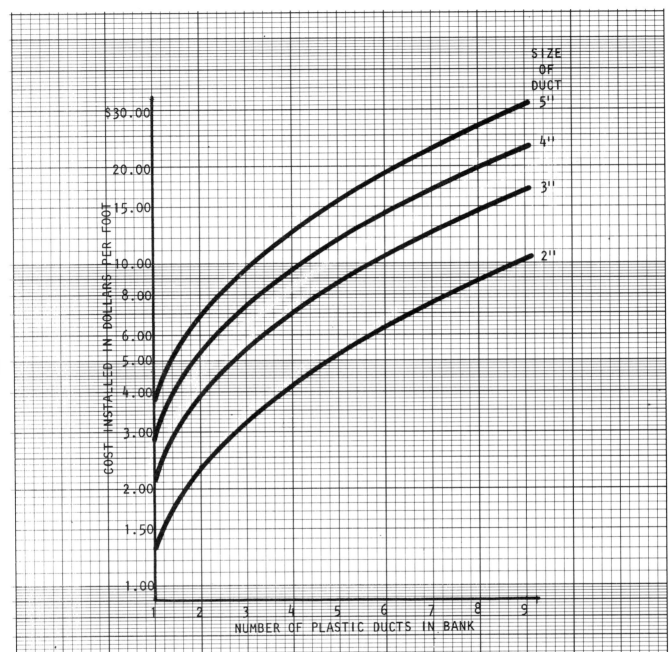

The costs shown for nonmetallic duct, earth cover, include the published contractors' book price for plastic duct of the sizes shown. Included are the spacers, cement, and #10 iron fish wire for each duct.

Labor is included for installation of the above; however, no trenching or backfilling is provided in the costs shown. See page L-2, Miscellaneous Section, for trenching and backfilling.

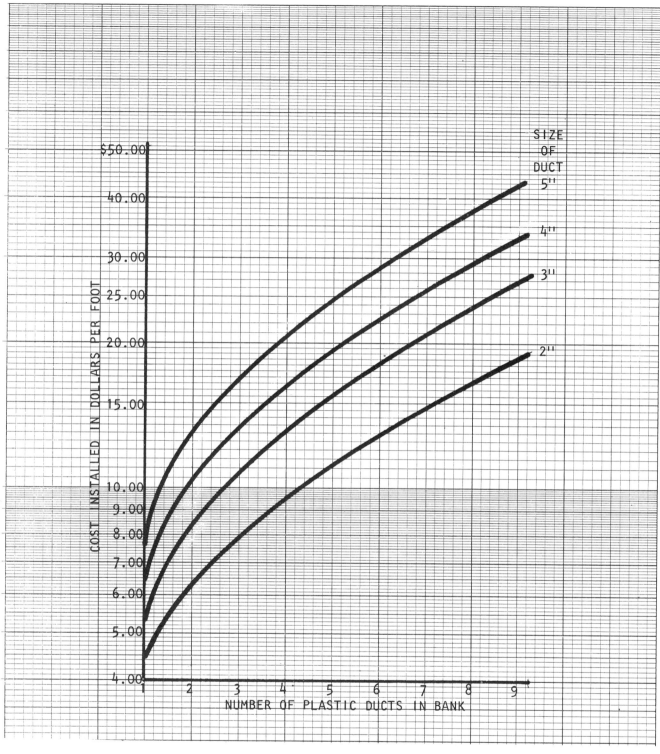

The costs shown for the nonmetallic duct bank, concrete-encased,
include the published contractors' book price for plastic duct of the
sizes shown. Including are the couplings, cement, plastic spacers on
5-foot centers for 2-inch separation in both directions, #10 iron fish
wire for each duct, and concrete to provide a 3-inch envelope around the
duct bank. Labor is included for the installation of the above; however,
no trenching or backfilling is provided in the costs shown. See page L-2,
Miscellaneous Section, for trenching and backfilling.

The costs shown for the nonmetallic duct bank, reinforced-concrete-encased, include the published contractors' book prices for plastic duct of the sizes shown. Included are the cement, plastic spacers on 5-foot centers for 2-inch separation in both directions, #10 iron fish wire for each duct, rebar, and concrete to provide a 3-inch envelope around the duct bank. Labor is provided for the installation of the above; however, no trenching or backfilling is provided in the costs shown. See page L-2, Miscellaneous Section, for trenching and backfilling.

COST OF NONMETALLIC DUCT BANK

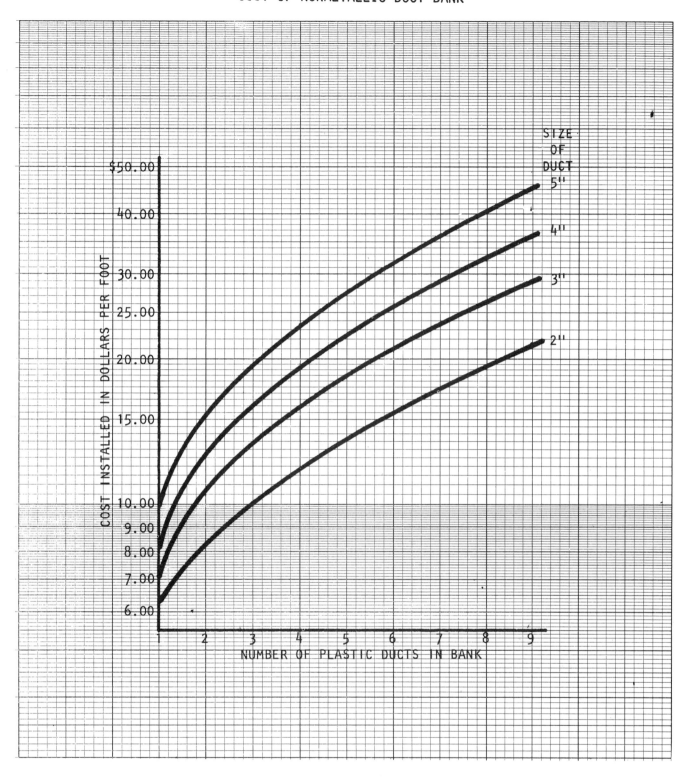

DUCT BANK - NONMETALLIC - REINFORCED CONCRETE

The costs shown for ladder tray and fittings consist of the contractors' published book prices for the tray and fittings shown. All tray and fittings shown here are 4 inches deep and have rung spacing of 9 inches on centers and are as manufactured by Square D Co.

Included: (material and labor)
1. Splice plates and bolts.
2. Two 36-inch hanger rods on 5-foot centers.
3. Rod-suspension assemblies appropriate to the tray.

Excluded:
1. Freight and cartage from factory.
2. Covers.

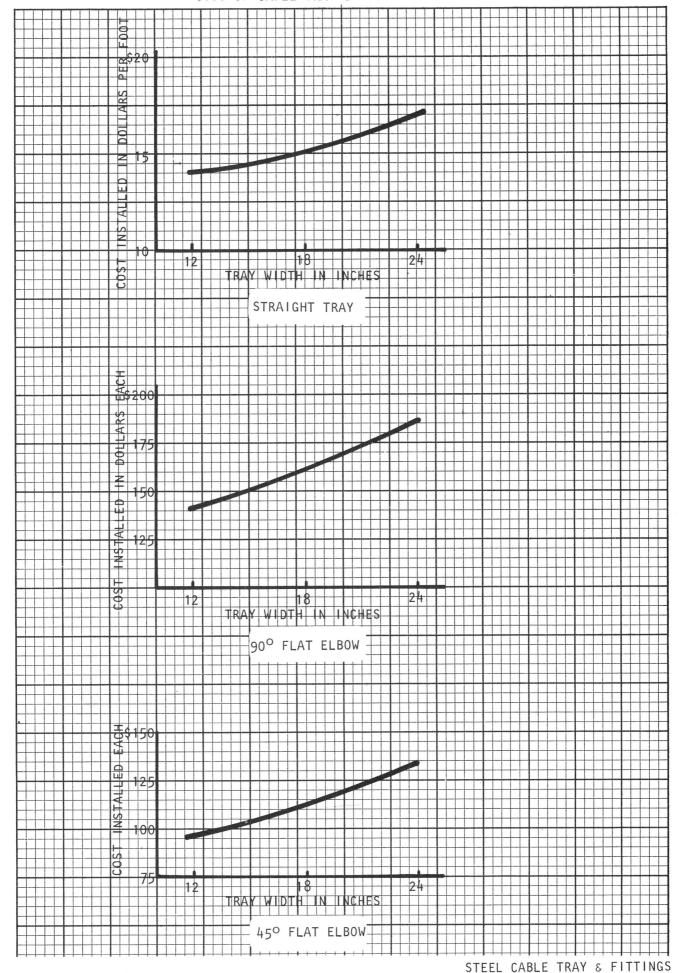

STRAIGHT TRAY

90° FLAT ELBOW

45° FLAT ELBOW

The costs shown for these reducers consist of the contractors' published book prices for the units shown and are as manufactured by Square D Co.

Included: (material and labor)
  Splice plates and necessary bolts for each unit.

Excluded:
  Freight and cartage from factory.
  Hanger rod on reducer.
  Covers.

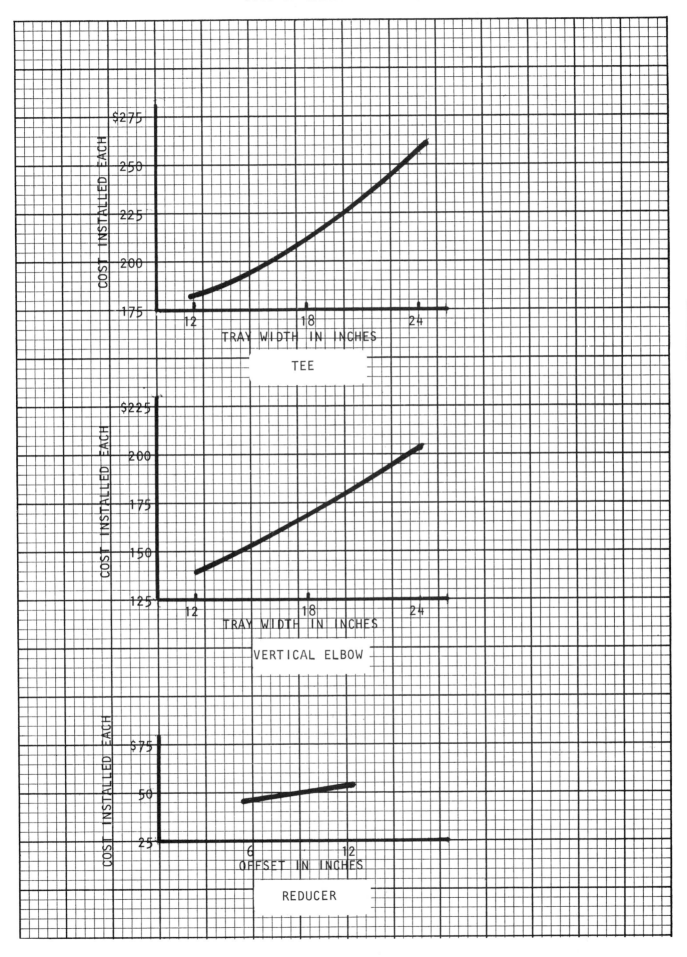

STEEL CABLE TRAY FITTINGS

Branch Circuits

CONDUITS—WITH WIRE

EMT with Copper TFF   F-1
EMT with Copper THWN/THHN   F-2
IMC or GRC with Copper TFF   F-3
PVC with Copper TFF or THHN   F-4

CONDUITS—EMPTY

EMT, ARC, GRC, PVC, FLEX, Liquidtight,
    PVC-Coated GRC, and IMC   F-5
Nonmetallic Tubing and FLEX.   F-6

CONDUIT FITTINGS

Terminals   F-7
Conduit Fittings   F-8
Erickson Couplings   E-4
Nipples   E-5

CONDUCTORS

Copper TFF, THHN   F-9
Thermostat/Annunciator   F-10
Nonmetallic Sheathed Cable—Copper   F-11
XLPE/USE—Direct Burial—Copper   F-12
Nonmetallic Sheathed Cable—UF—Copper—Direct Burial   F-13
Armored Cable—AC—Copper Only   F-14
Bare Copper Wire   D-2
Grounding Conductor—Insulated   E-16

MINERAL-INSULATED CABLE

Mineral-Insulated Cable—Nonjacketed   F-15
Mineral-Insulated Cable—Jacketed   F-16
Terminations for Mineral-Insulated Cable   F-17

OUTLETS

Fixture Outlets—Ceiling and Wall Type   F-18
Receptacles—Duplex, Single, Weatherproof, Dryer, Range   F-19
Receptacles—Built-in Ground Fault Interrupter   F-20
Clock, Floor, Telephone   F-21
Switches—Wall and Door   F-22
Fire-Rated Floor Fittings   F-23

F

LOW-VOLTAGE REMOTE CONTROL SYSTEM—SEE
CONTROL EQUIPMENT

APPLIANCE CONNECTIONS   F-24

MOTOR TERMINAL CONNECTIONS

115-, 200-, and 230-Volt Motors   F-25
460-Volt Motors   F-26

SURFACE RACEWAYS

Channel Support and Raceway System—Unistrut   F-27
Wiremold Raceways   F-28
Wiremold Fittings, Devices, and Plugmold   F-29
Telepower Poles   F-30

UNDERCARPET WIRING SYSTEM

Flat Conductor Cable   F-31
Wiring System Components   F-32

UNDERFLOOR RACEWAY SYSTEM

Standard and Super Metal Underfloor Duct   F-33
Underfloor Duct Junction Boxes and Fittings   F-34

TRENCH DUCT SYSTEM

Trench Duct   F-35
Trench Duct Fittings   F-36, F-37

LIGHTING DUCT SYSTEM

Lighting Duct Sections and Fittings   F-38
Lighting Duct Terminals and Receptacles   F-39

The costs shown for EMT conduit and TFF wire include the published contractors' book price and labor for installation of both conduit and wire. The graph complies with the 1984 National Electrical Code, Chapter 9, Table 2.

COST OF EMT WITH WIRE

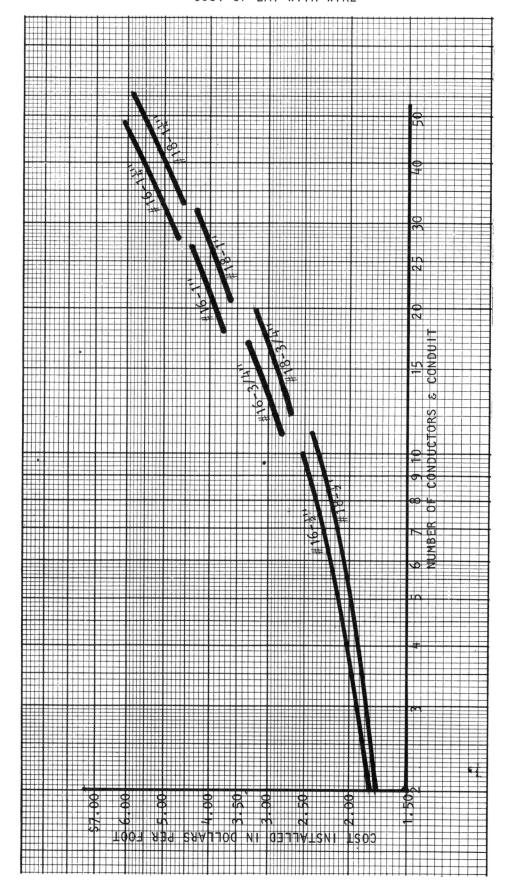

EMT CONDUIT WITH COPPER TFF CONDUCTORS

F

F-1

The costs shown for the EMT conduit and THWN/THHN wire include the published contractors' book price and labor for installation of both conduit and wire. The graph complies with the 1984 National Electrical Code, Chapter 9, Table 3B.

COST OF EMT WITH WIRE

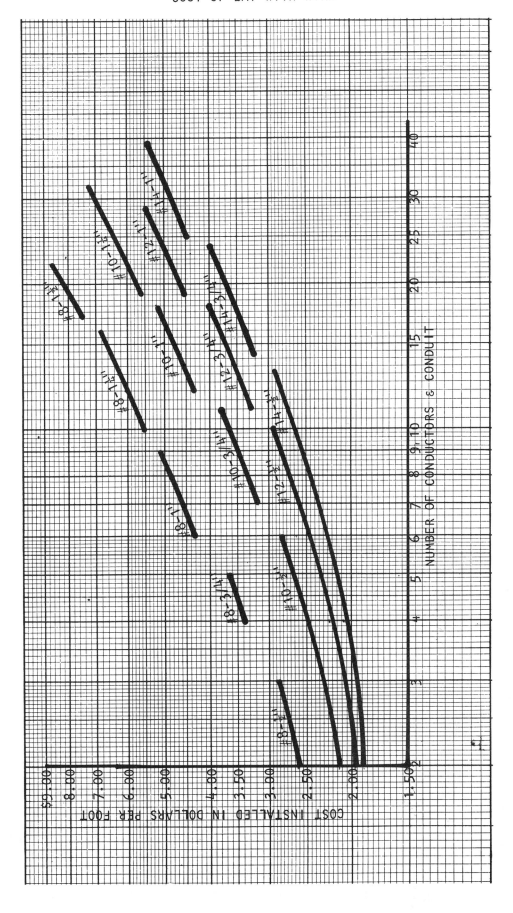

EMT CONDUIT WITH COPPER THWN/THHN CONDUCTORS

F

F-2

The costs shown for the IMC and GRC conduit with TFF wire include the published contractors' book price and labor for installation of both conduit and wire. The graph complies with the 1984 National Electrical Code, Chapter 9, Table 2.

COST OF IMC or GRC CONDUIT & WIRE

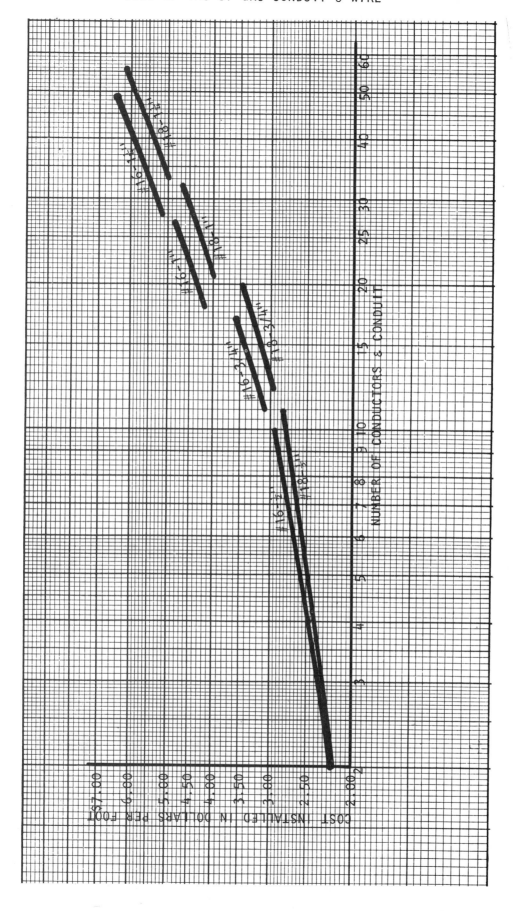

IMC or GRC CONDUIT WITH COPPER TFF CONDUCTORS

F

F-3

The costs shown for PVC conduit with TFF or THHN wire include the published contractors' book price and the labor for installation of both conduit and wire.

COST OF PVC CONDUIT & WIRE

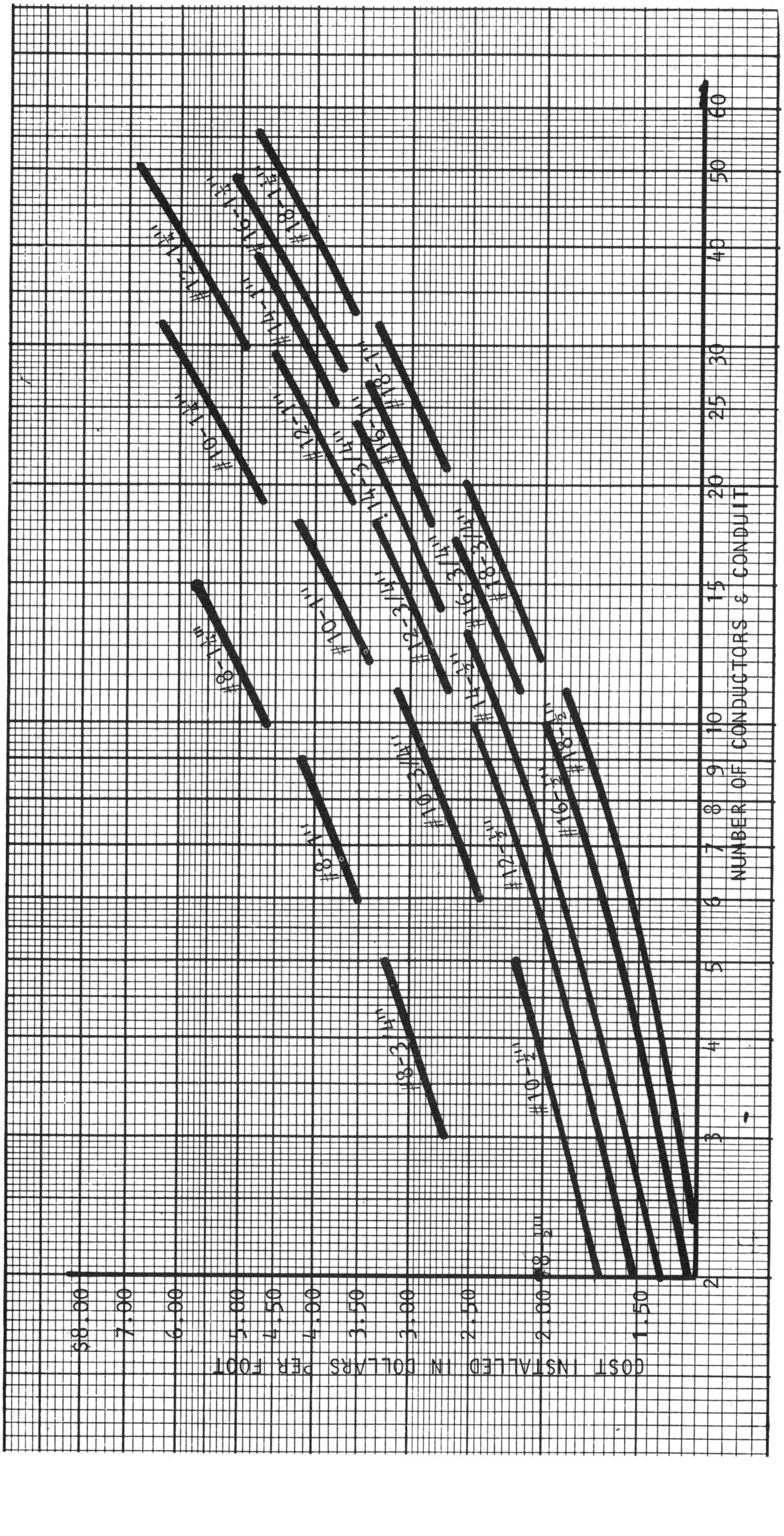

PVC CONDUIT WITH COPPER TFF OR THHN CONDUCTORS

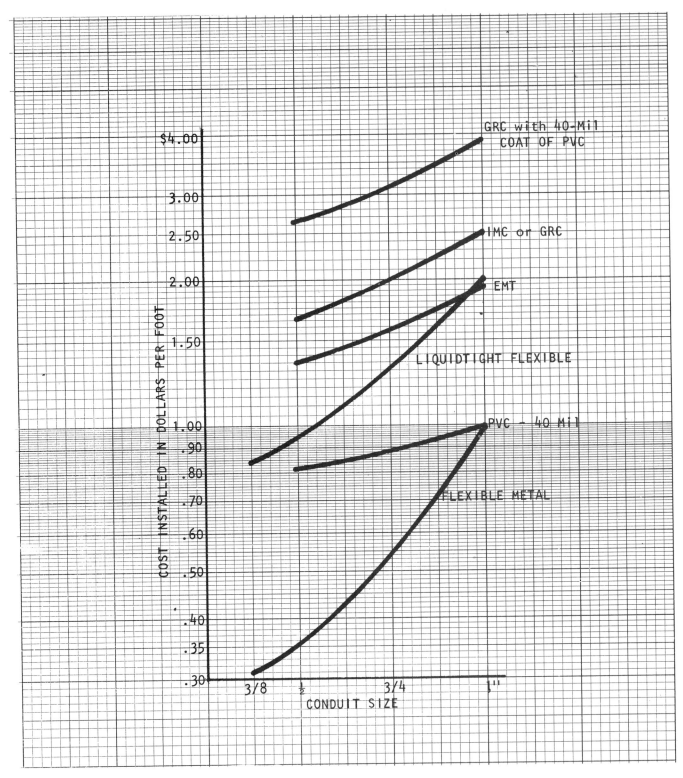

The costs shown for the branch circuit raceways include the published contractors' book prices and the labor required for the installation.

F-5

COST OF NONMETALLIC BRANCH CIRCUIT CONDUITS

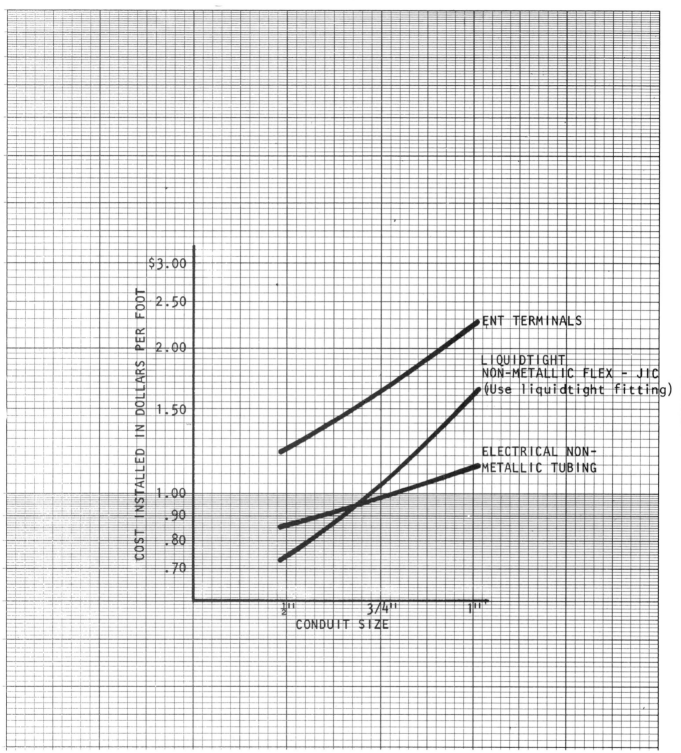

ENT TERMINALS

LIQUIDTIGHT
NON-METALLIC FLEX - JIC
(Use liquidtight fitting)

ELECTRICAL NON-
METALLIC TUBING

COST INSTALLED IN DOLLARS PER FOOT

$3.00
2.50
2.00
1.50
1.00
.90
.80
.70

CONDUIT SIZE

1/2"   3/4"   1"

F

F-6

The costs shown for these branch circuit conduits include the published
contractors' book price and the labor for installation. Electrical
nonmeticallic tubing is in the new Article 331 of the 1984 National
Electrical Code.

F-7

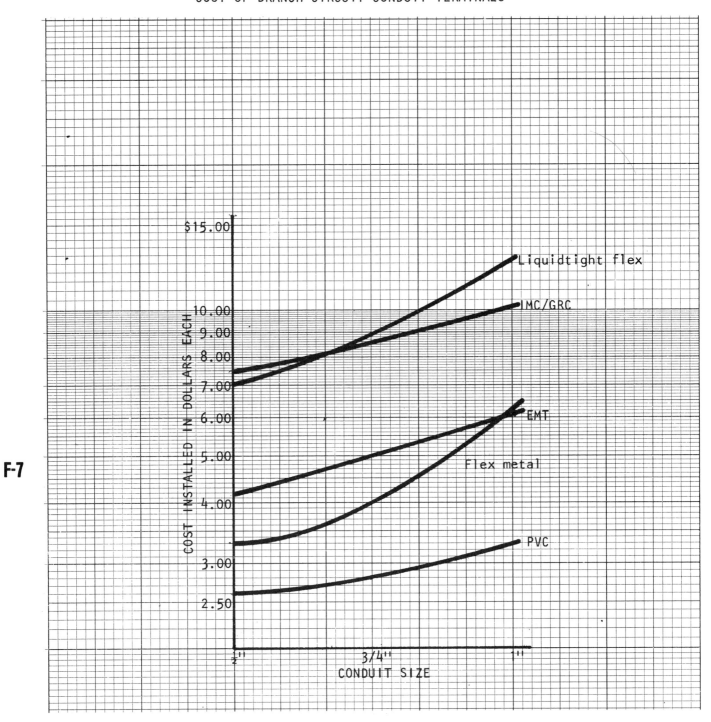

The costs shown for conduit terminals include the published contractors' book price for the type of terminal required, such as locknuts and insulated metallic bushings for IMC and GRC, insulated throat connectors for EMT, plastic connectors for PVC, and connectors for FLEX. It further includes the labor for preparing the end of the conduit either by cutting and threading or by cutting and glueing as needed to satisfy the requirements of the conduit.

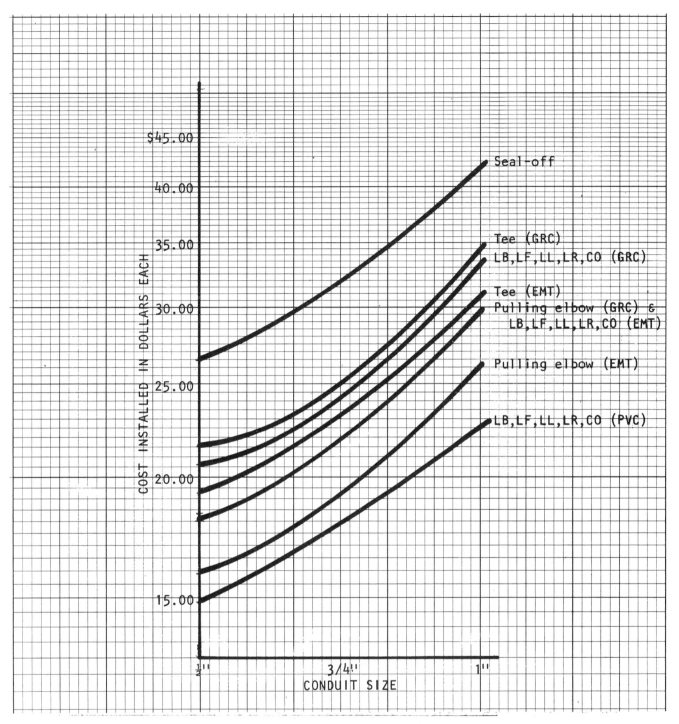

The cost shown for the conduit fittings includes the published contractors'
book price for the various fittings shown of the gasketed type to suit the
kind of conduit required. These fittings are basically LBs, LLs, LRs,
and LFs. The cost also includes the labor of preparing the end of the
conduit and installing the fitting. Seal-offs are considered to be used
with GRC in hazardous applications and include the installation of
sealing compound as required.

F

F-8

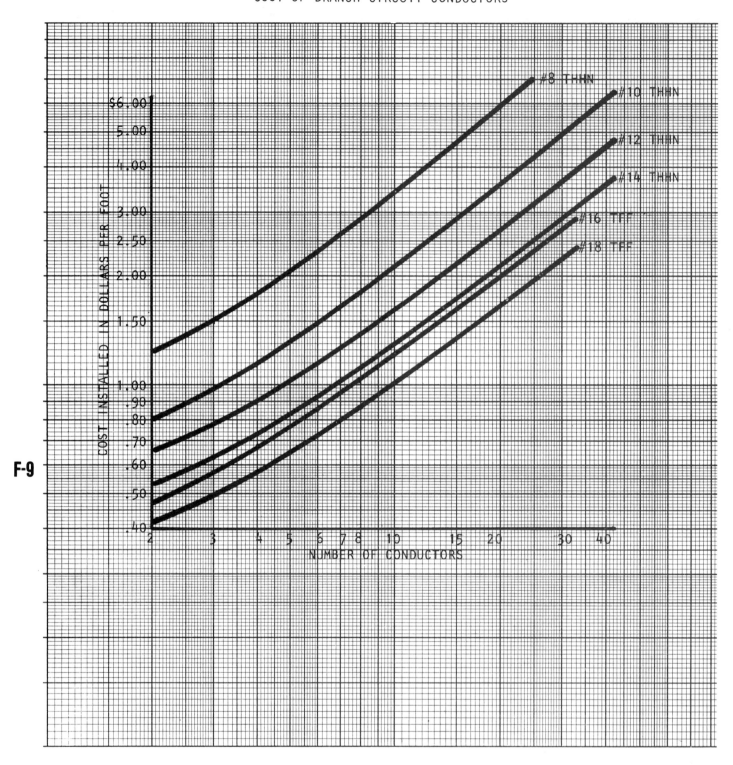

F-9

The cost shown for the branch circuit conductor includes the published contractors' book prices and the labor of installing the conductors in the raceways.

COST OF THERMOSTAT/ANNUNCIATOR CABLE

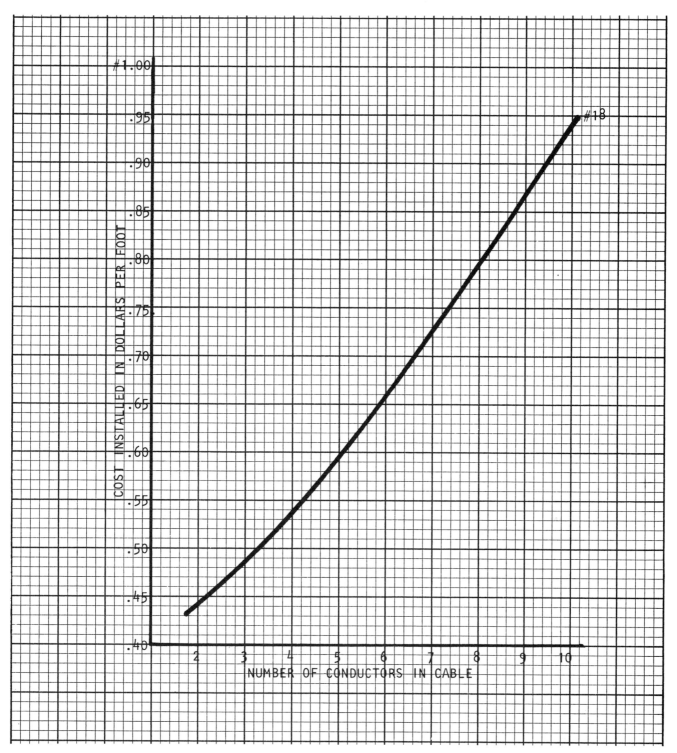

The cost shown for the thermostat/annunciator cable is for plastic insulation and plastic jacket and includes the published contractors' book price and the labor for installing the cable in conduit as one conductor.

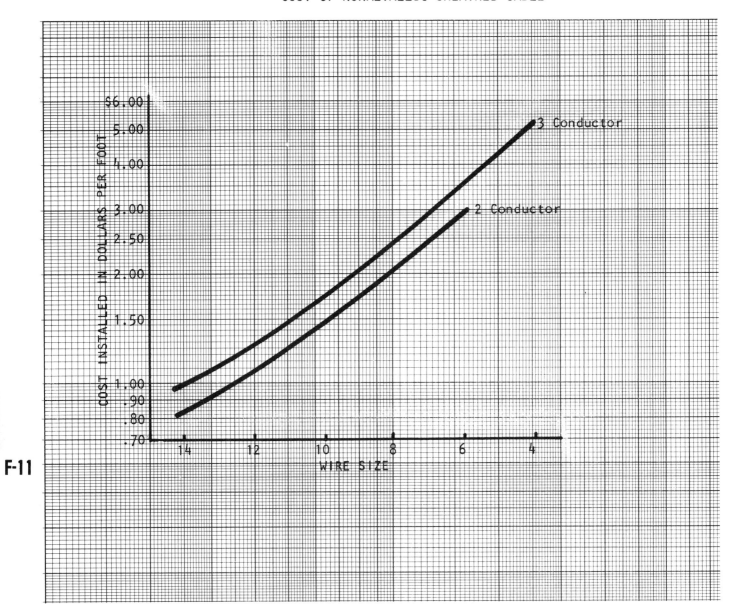

F-11

The costs shown for the nonmetallic sheathed cable include the published contractors' book prices and also the labor of installing the NMSC on a wood surface or joists supported by staples on 4-foot centers. The grounding conductor is full-size.

NONMETALLIC SHEATHED CABLE WITH GROUND - COPPER

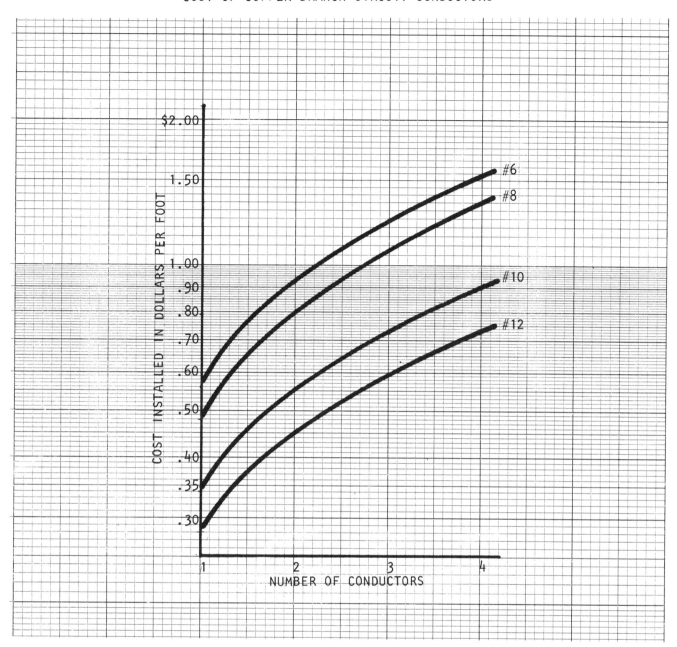

The costs shown for the direct-burial conductors include the published contractors' book price for the kind of insulation shown and the labor for installing in an existing open trench. Trenching and backfilling are not included. See Miscellaneous Section.

F

F-12

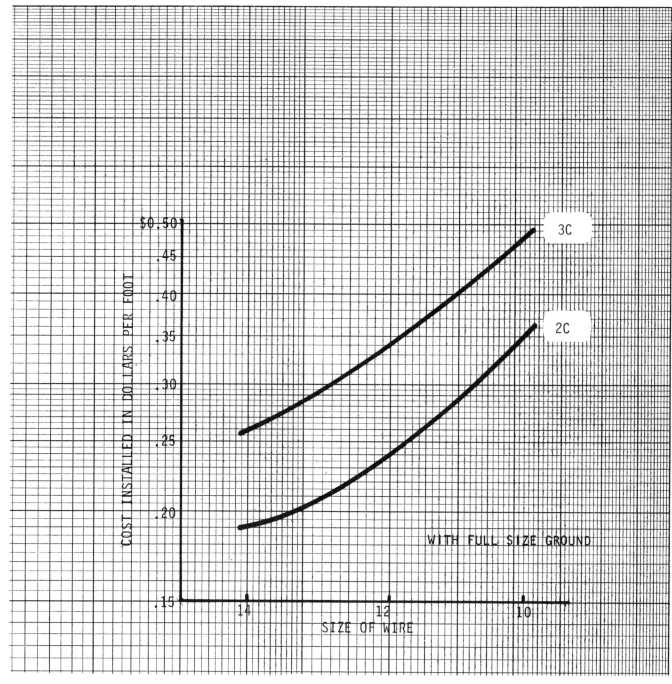

F-13

The cost shown for the underground feeder (UF) cable is based upon the contractors' book prices for the number of conductors shown, including the grounding conductor. Labor is based upon installing in an existing trench. Trenching and backfilling costs are not included. See page L-2, Miscellaneous Section, for these costs. The grounding conductor is full-size.

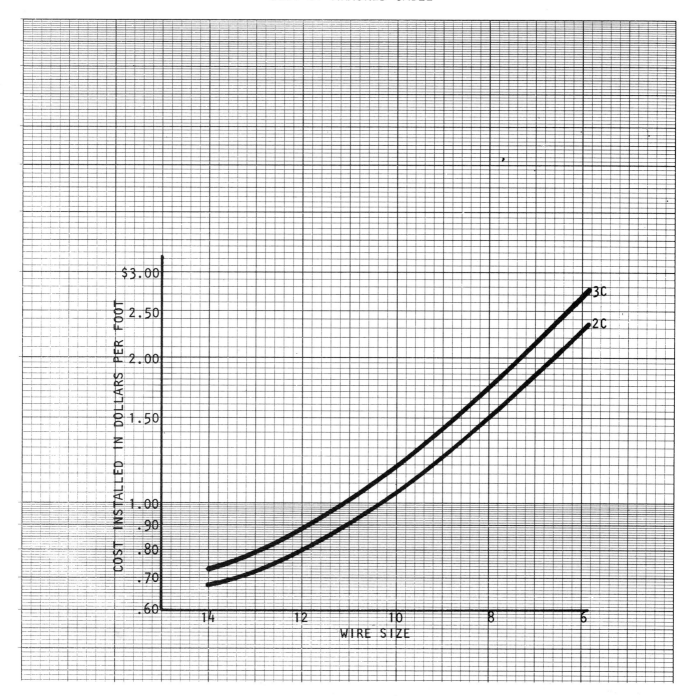

The costs shown for the armored cable include the published contractors'
book price and the labor for installation in bored wood joists. Conductors
are copper only.

The cost shown for the mineral-insulated cable consists of the contractors' published book prices and installation on the surface with the number and size of conductors shown.

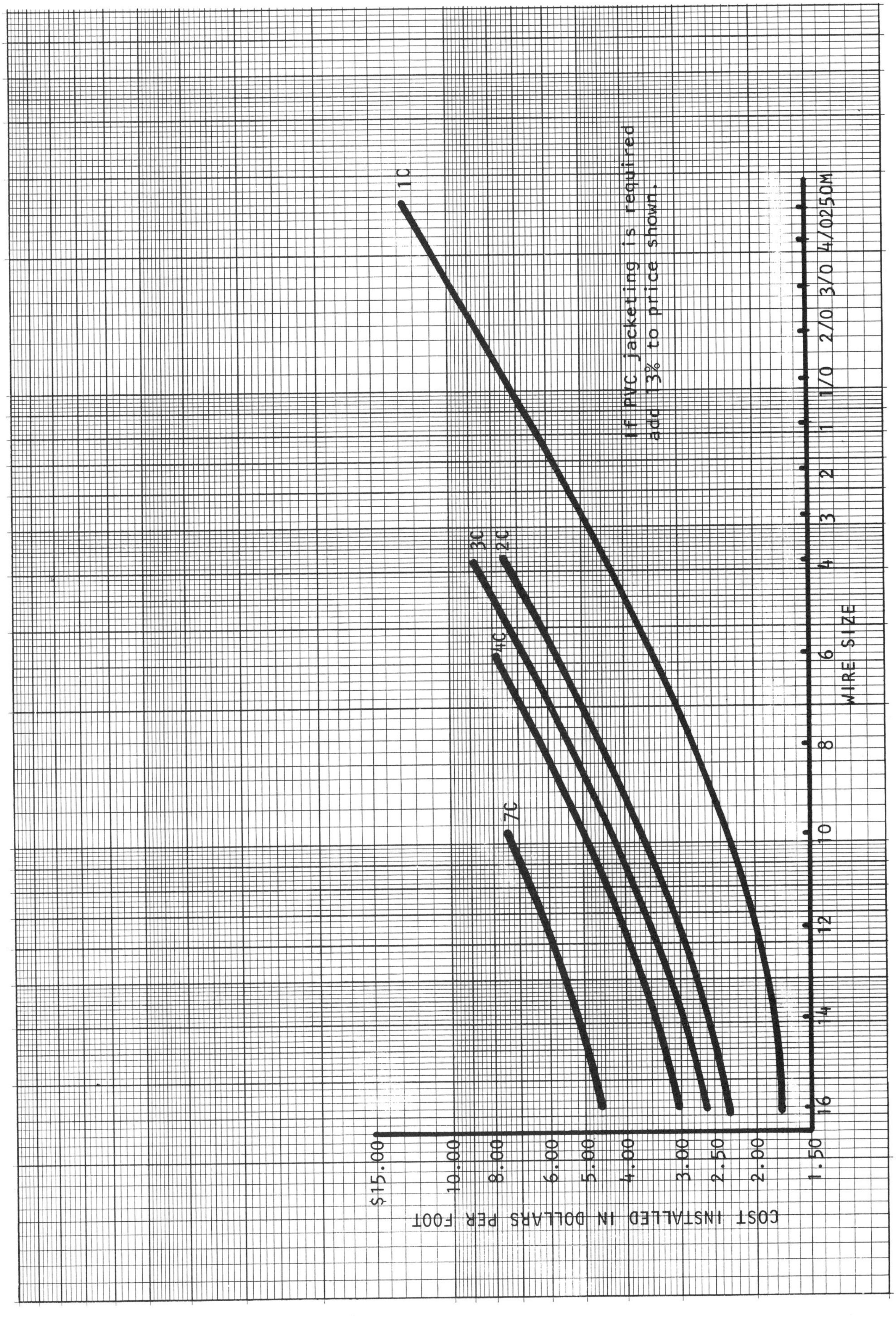

COST OF MINERAL-INSULATED WIRING CABLE (NORMAL DUTY)

F

F-15

The chart plots **COST INSTALLED IN DOLLARS PER FOOT** (vertical axis, values $15.00, 10.00, 8.00, 6.00, 5.00, 4.00, 3.00, 2.50, 2.00, 2.00, 1.50) against **WIRE SIZE** (horizontal axis, values 16, 14, 12, 10, 8, 6, 4, 3, 2, 1, 1/0, 2/0, 3/0, 4/0250M).

Curves labeled: 1C, 2C, 3C, 4C, 7C

If PVC jacketing is required add 3% to price shown.

MINERAL-INSULATED WIRING CABLE

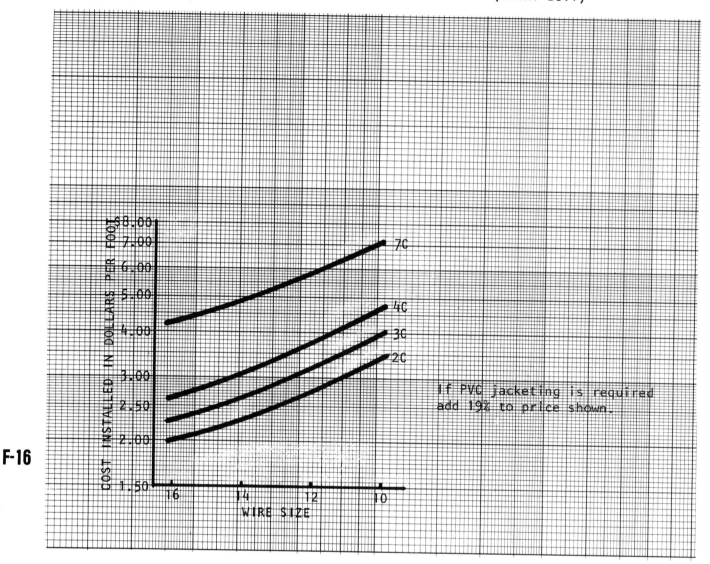

F-16

The costs shown for the modified mineral-insulated wiring cable consist of the contractors' published book prices and the labor for installation on the surface. The cable is designed primarily for lighting, control circuits, and light-duty applications.

COST OF MINERAL-INSULATED CABLE TERMINATIONS

COST INSTALLED IN DOLLARS EACH

FITTING SIZE	WIRE SIZE	NUMBER OF CONDUCTORS				
		1	2	3	4	7
1/2"	16	$12.50	$14.00	$15.50	$17.50	$24.75
	14	13.00	14.50	16.25	21.50	26.00
	12	14.00	15.25	20.50	22.75	27.25
	10	15.00	19.75	21.75	24.25	39.00
	8	16.25	21.50	23.75	26.25	
	6	18.00	23.25	26.00	38.25	
	4	20.00	34.25	38.75		
3/4"	3	24.75				
	2	26.00				
	1	27.50				
	1/0	29.25				
	2/0	41.00				
	3/0	43.00				
1"	4/0	45.00				
	250MCM	47.25				

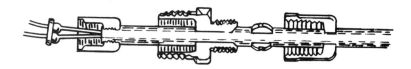

The cost shown for the mineral-insulated cable terminations consists of the contractors' published book prices and labor for installation in accordance with the manufacturer's instruction. Some special tools are required.

The cost shown for ceiling outlets mounted flush in wood frame construction includes a 4-inch octagon box with plaster ring mounted to an adjustable bar hanger with a fixture stud and two connectors of the type required.

Recessed above the ceiling, the outlet is designed for use with a recessed fixture and consists of a 4-inch square box and blank cover, bar hanger, two connectors for cable or conduit, a 4-foot piece of 3/8-inch flexible metal conduit with two connectors, and three 6-foot lengths of #12 AF wire.

Flush pan or deck mounting is in concrete and contains a 4-inch concrete ring, plate with fixture stud, and connectors for the type of conduit required.

COST OF FIXTURE OUTLETS

CEILING OUTLETS

Type Terminal	Type of Construction					
	Flush Mounting			Surface Mounting		
	Wood Frame	Recessed Above Susp. Clg.	Flush Pan or Deck	Wood	Steel	Concrete
Non-metallic Sheathed Cable	$17.50	- - - -	- - -	- - -	- - -	- - -
Armored Cable	23.00	- - - -	- - -	- - -	- - -	- - -
EMT	27.00	$33.75	$23.75	$21.75	$22.00	$26.75
GRC	- - -	40.75	30.75	31.50	32.00	36.50

WALL OUTLETS

Type Terminal	Type of Construction							
	Flush Mounting				Surface Mounting			
	Wood Frame	Metal Studs	Block Masonry	Reinf. Concrete	Wood	Steel	Concrete	Block Masonry
Non-metallic Sheathed cable	$8.75	$10.25	- - - -	- - - -	- - -	- - -	- - - -	- - -
Armored Cable	12.50	14.50	- - - -	- - - -	- - -	- - -	- - - -	- - -
EMT	22.50	28.25	$23.00	$38.00	$21.75	$21.75	$26.75	$26.75
GRC	30.00	38.25	30.00	44.00	31.50	31.00	36.50	36.75

Receptacles shown on the facing page include much the same materials as described on page F-18, but, in addition, the duplex receptacles contain 15-ampere-rated specification-grade devices unless otherwise indicated and painted metal wall plates. Outlets served by nonmetallic sheathed cable or armored cable have standard-grade devices. Surface-mounted 30- and 50-ampere outlets are of the molded-plastic dryer and range type.

COST OF RECEPTACLES

RECEPTACLES

	TYPE OF CONSTRUCTION								
	FLUSH MOUNTING					SURFACE MOUNTING			
TYPE TERMINAL	WOOD FRAME	METAL STUDS	BLOCK MASONRY	REINF. CONCRETE	CONDUIT SUPPORTED	WOOD	STEEL	CONCRETE	BLOCK MASONRY
DUPLEX									
N.M.S.C.	$29.65	$28.50	—	—	—	$20.75	—	$26.75	$25.00
Armored Cable	30.00	34.00	—	—	—	—	—	—	—
EMT	40.00	41.00	$39.75	$55.50	$45.50	36.00	$35.50	38.25	38.25
GRC	46.75	47.25	46.75	62.00	47.25	45.75	45.50	48.50	48.50
WEATHERPROOF DUPLEX									
N.M.S.C.	37.50	36.25	—	—	—	—	—	—	—
Armored Cable	37.75	41.85	—	—	—	—	—	—	—
EMT	48.00	49.00	47.50	—	59.75	59.75	59.75	59.75	59.75
GRC	55.00	55.25	54.50	—	61.75	61.75	61.75	61.75	61.75
SINGLE									
20A-3W-EMT	42.50	44.00	42.25	58.25	48.25	39.00	38.75	41.50	41.50
GRC	49.50	50.00	49.50	64.50	50.25	48.50	48.25	51.25	51.25
30A-3W-RMX/BX	46.00	47.00	45.00	—	—	40.50	—	—	42.75
EMT	46.25	47.25	46.25	62.00	64.00	40.50	40.00	42.75	42.75
GRC	53.25	53.50	52.75	68.00	65.75	50.25	50.00	52.50	52.50
50A-3W-RMX/BX	54.25	53.50	50.00	—	—	46.25	—	—	49.00
EMT	54.25	53.50	52.50	68.00	70.50	46.25	—	49.00	49.00
GRC	61.75	60.75	60.00	75.25	74.25	57.75	—	60.50	60.50

DUPLEX, WEATHERPROOF DUPLEX, AND SINGLE RECEPTACLES

F

F-19

GROUND-FAULT CIRCUIT-INTERRUPTER		
TYPE	INSTALLED COST	
	15 Amp.	20 Amp.
INDOOR	$80.25	$83.50
WEATHERPROOF	88.50	91.75

Costs shown include outlet box and single-gang ring mounted in plaster wall and GFI receptacle with painted metal (interior) wall plate.

F-20

COST OF CLOCK, FLOOR, AND TELEPHONE OUTLETS

CLOCK OUTLETS

Type Terminal	Type of Construction			
	Flush Mounting			
	Wood Frame	Metal Studs	Hollow Masonry	Reinf. Concrete
EMT	$46.25	$46.75	$45.50	$61.50
GRC	52.50	52.50	52.00	67.25

FLOOR OUTLETS

Type Terminal	Type of Construction	
	Flush Mounting	
	Wood Frame	Concrete Slab
EMT	$76.25	$72.50
GRC	78.00	74.50

TELEPHONE OUTLETS

Type Terminal	Type of Construction					Surface Mounting
	Flush Mounting					Surface Mounting
	Metal Studs	Block Masonry	Reinf. Concrete	Concrete Slab Floor	Wood Floor	Block Masonry
EMT	$26.00	$24.75	$40.50	$77.00	$65.75	$22.75

The clock, floor, and telephone outlets are specification-grade devices installed as described on page F-18. Those outlets mounted in the slab are pressed-steel floor boxes of the adjustable type and concrete-tight. The receptacles are flush in the floor.

F

F-2

The switches shown in this table include much the same materials as described on page F-18 but, in addition, contain a 15-ampere-rated specification-grade quiet type of switch with a painted metal wall plate.

All devices priced in this table are of specification-grade quality. No residential-grade or standard-grade devices are used in pricing.

Type of Construction								
	Surface Mounting				Flush Mounting			
Type Terminal	Block Masonry	Concrete	Steel	Wood	Reinf. Concrete	Block Masonry	Metal Studs	Wood Frame
SINGLE POLE								
N.M.S.C.	—	—	—	—	—	—	$23.00	$21.75
Armored Cable	—	—	—	—	—	—	27.25	25.25
EMT	$33.50	$33.50	$31.00	$31.25	$49.00	$36.00	38.50	33.00
GRC	43.25	43.25	40.25	40.50	54.75	42.25	48.00	40.00
3-WAY SWITCH								
N.M.S.C.	—	—	—	—	—	—	28.00	26.75
Armored Cable	—	—	—	—	—	—	52.25	30.25
EMT	38.75	38.75	35.75	36.25	54.00	41.00	43.50	38.25
GRC	48.25	48.25	45.25	45.50	59.75	47.25	52.75	45.00
DOOR SWITCH								
N.M.S.C.	—	—	—	—	—	—	—	74.75
Armored Cable	—	—	—	—	—	—	—	76.75
Flex metal	—	—	—	—	—	—	—	79.25
EMT	—	—	—	—	—	—	—	80.75
GRC	—	—	—	—	—	—	—	87.50

F

F-22

Concrete Slab Thickness	Installed Cost		
	Telephone	Duplex Receptacle	Tele. & Rec.
4"	$85.75	$98.50	$148.25
6"	89.00	100.75	153.75
8"	91.75	103.25	160.25

Costs of fittings include the published contractors' book price for the assembly including the combination power–telephone-service fitting and the cost of core-drilling the concrete floor slab. Abandon plate includes the labor expense of fitting removal.

F-23

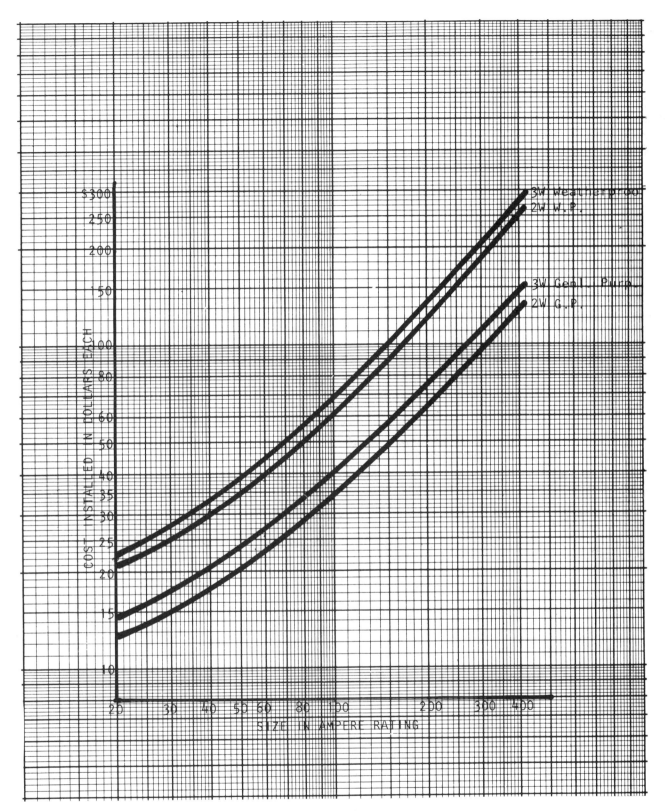

The costs shown for an appliance connection consist of 3 feet of flexible
metal conduit for general-purpose use, or liquidtight flexible metal
conduit for weatherproof use, with two connectors of the proper type,
4 feet of two or three conductors, and wire terminals, plus a grounding
conductor sized in accordance with NEC Article 250–95. Labor is
provided to make the final installation. Add for a safety switch if required.

F

F-24

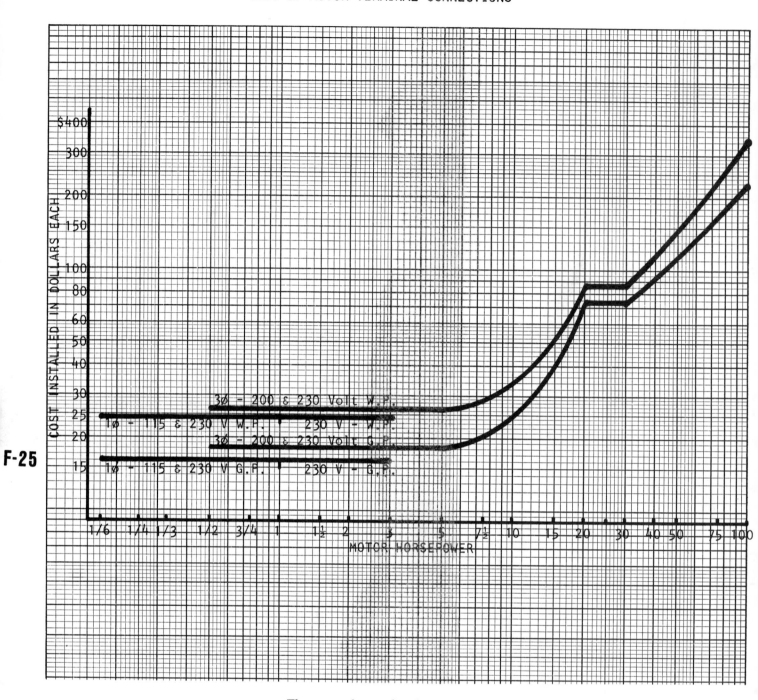

F-25

The costs shown for the motor terminal connections consist of 3 feet of flexible metal conduit, two conduit connectors, 4 feet of three conductors, and three lugs for three-phase connections. The weatherproof connection has liquidtight flexible metal conduit with the proper connectors and conductors. Up to 3 horsepower the cost is the same as shown by the straight horizontal lines. Labor is provided to make the final installation. Conductor and conduit sizes comply with the National Electrical Code.

These costs do not include a motor disconnect switch. If you have priced a magnetic starter, do not include these costs as they have been included with the starter.

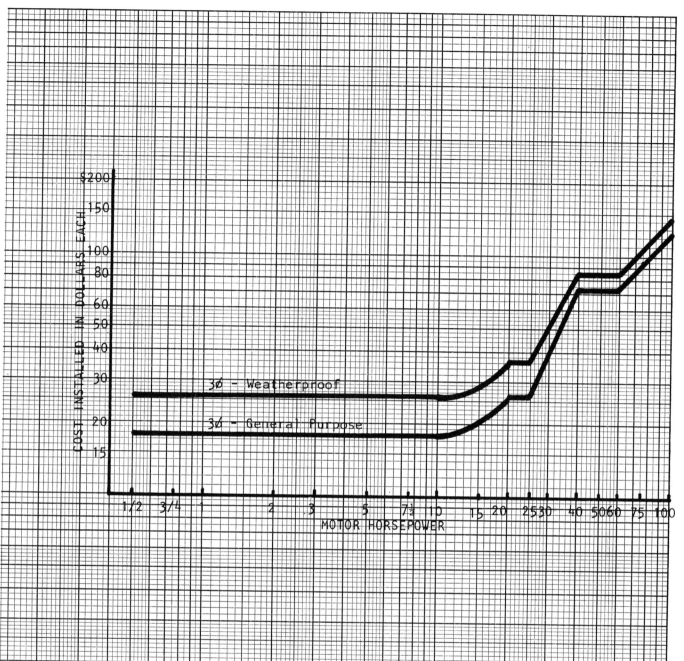

The costs shown for the motor terminal connections consist of 3 feet of flexible metal conduit, two conduit connectors, 4 feet of three conductors, and three lugs for three-phase connections. The weatherproof connection has liquidtight flexible metal conduit with the proper connectors and conductors. Up to 10 horsepower the cost is the same as shown by the straight horizontal lines. Labor is provided to make the final installation. Conductor and conduit sizes comply with the National Electrical Code.

These costs do not include a motor disconnect switch. If you have priced a magnetic starter, do not include these costs since they have been included.

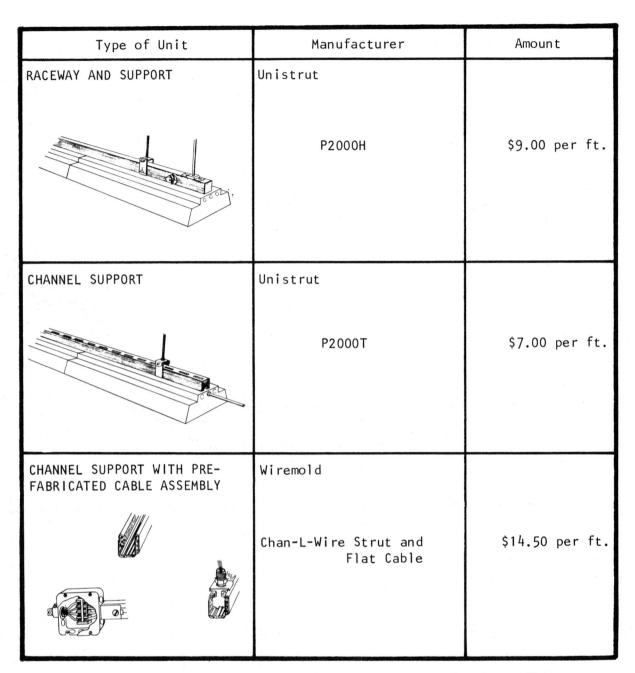

Type of Unit	Manufacturer	Amount
RACEWAY AND SUPPORT	Unistrut P2000H	$9.00 per ft.
CHANNEL SUPPORT	Unistrut P2000T	$7.00 per ft.
CHANNEL SUPPORT WITH PRE-FABRICATED CABLE ASSEMBLY	Wiremold Chan-L-Wire Strut and Flat Cable	$14.50 per ft.

F-27

The raceway and support systems shown are based upon Unistrut P-2000-H closure strip on top of the raceway, stud nuts for supporting fixtures, 36-inch hanger rod and fittings every 10 feet, connectors, end caps, etc., but no wire or fixtures. Labor is included for installing raceways but not fixtures or conductors.

The channel support system is not a raceway system and has a slotted channel. Included are the slotted channel, hangers at 10-feet intervals, and T bolts for supporting fixtures, but no fixtures are provided.

The cost of the channel support with cable assembly is based on typical runs of 100 feet and includes Wiremold Strut special splice plates, terminal box, insulating end cap, prefabricated cable assembly installed, 25 power tap fixture hangers, and 25 nonpower fixture hangers. Lighting-fixture-mounting labor is not included.

The cost shown for Wiremold raceways and fittings consists of the published contractors' book prices and the labor for installation. These raceways and fittings are considered to be mounted on concrete using Rawl plugs and wood screws.

The costs shown for the Wiremold devices consist of the published contractors' book prices and the labor for installation. These devices are considered to be mounted on concrete using Rawl plugs and wood screws.

Plugmold

   a. *20G and 22G Series: 2 wire, 1 circuit, where the Plugmold base serves as the grounding conductor.*

   b. *20GA and 22GA Series: 3 wire, 2 circuit, with outlets wired alternately. Plugmold base serves as grounding conductor.*

   c. *20DGA and 22DGA Series: 3-wire duplex outlets, 2-circuit grounding. Each duplex is wired alternately. Raceway serves as grounding conductor.*

COST OF SURFACE RACEWAY FITTINGS

COST INSTALLED IN DOLLARS EACH

	200	500	700	1500	2000	2100	2200	3000	4000	6000
Internal elbow	$13.00	$12.50	$12.50	$18.25	$13.50	$16.25	$17.00	$16.00	$26.00	$35.50
External elbow	11.00	12.50	12.50	18.50	11.00	15.75	14.50	18.75	32.50	35.50
Flat elbow	12.50	12.50	12.50	18.25	16.25	16.50	25.00	24.25	27.50	33.00
Fixture of Extension outlet	21.75	21.50	23.75	-----	-----	-----	-----	-----	-----	-----
Duplex receptacle	29.25	34.25	34.25	31.75	-----	17.75	-----	21.25	25.50	-----
Wall switch	27.00	32.50	32.50	-----	20.00	21.75	-----	17.75	28.00	-----
Telephone outlet	-----	-----	-----	30.00	-----	-----	-----	-----	-----	-----

COST OF PLUGMOLD MULTIOUTLET SYSTEM INSTALLED IN DOLLARS PFR FOOT

Outlet spacing	Single circuit - 2 wire Grounding type		Two circuit - 3 wire Grounding type		Two circuit - 3w- grounding	
	20G	22G	20GA	22GA	Duplex 20DGA	Duplex 22DGA
6"	$8.00	-----	-----	-----	-----	-----
12"	6.50	-----	6.75	-----	-----	-----
18"	5.50	6.25	6.00	-----	-----	-----
30"	5.00	6.00	5.50	5.75	5.75	6.50
60"	4.75	5.50	5.00	5.50	5.00	5.75

F

F-29

Type of Unit			
Manufacturer	Wiremold 1 1/4 x 1 3/4" Type 21TP-2 / 21TP-2G4 2 Outlets / 4 Outlets	Wiremold 2 3/4" x 1 7/16" Type 30TP-2 / 30TP-3 2 Duplex / 4 Outlets	Wiremold 2" x 2" Type ALTP-2 4 Outlets
Cost Installed	$121.50 $133.00	$143.50 $139.00	$177.00

F-30

The 21TP-2 telepower pole is essentially two sections of 2100 Wiremold back to back—one raceway for telephone and one for power with two grounding-type single outlets.

The 30TP telepower pole is essentially one section of 3000 Wiremold with two compartments. One will handle a 100-pair telephone cable for a call director. The other compartment handles the power wiring. Both the power and telephone wiring are fed down from above. The power side has two grounding-type duplex receptacles near the base.

The ALTP-2 is the deluxe telepower pole made from extruded 2-inch-square brushed aluminum with two separate compartments for the communication and power wiring. Various finishes are available. The power side has four single grounded-type receptacles. The communication side will adequately handle a 100-pair telephone cable. Power and communication cables feed down from the top.

The costs of the above telepower poles include the published contractors' book price and the labor required to assemble, wire, and install the pole to a grid ceiling and run 4 feet of flex metal conduit to an existing power junction box. All poles have an adjustable floor gripper pad.

Description		Cost installed in dollars per foot	
Undercarpet tile power cable		Slab on grade	Slab above grade
3-Conductor Cable. Available in 10 or 12 AWG, each of three colors.	Prices include steel tape 3 - #12 3 - #10	$7.55/Ft. 7.95/Ft.	$6.75/Ft. 7.15/Ft.
4-Conductor Cable Available in 10 or 12 AWG.	4 - #12 4 - #10	9.55/Ft. 9.95/Ft.	8.80/Ft. 9.20/Ft.
5-Conductor Cable Available in 10 or 12 AWG.	5 - #12 5 - #10	10.50/Ft. 10.90/Ft.	9.75/Ft. 10.15/Ft.

The cost of the flat conductor cable includes the published contractors' book price and the labor required to install the cable according to the manufacturer's recommendations. The "slab-on-grade" installation requires an additional PVC ribbon to protect the cable from the moisture below. The equipment shown is manufactured by Thomas & Betts Corp. and is covered under Article 328 of the 1984 National Electrical Code.

F

F-31

UNDERCARPET-TYPE FLAT CONDUCTOR CABLE

Description		Installed Cost
Transition Box (Round to flat cable) Transition boxes for standard stud or thin stud mounting.	Includes recessed box, terminal block, strain relief and transition connectors. 3 conductor cable 4 conductor cable 5 conductor cable	$72.00 76.00 80.25
Tee or Cross Connection Insulator Kit for Tap and Splice Connector Kit includes one each: top and bottom insulators, adhesive film for between cables, and end-cap tape.	Includes tap and splice connectors. 3 conductor cable 4 conductor cable 5 conductor cable	33.75 36.25 38.75
Low-Profile Pedestals Single Duplex	Includes base, receptacle, cover, & misc. hardware 15 amp. receptacle 20 amp. receptacle	61.50 63.75
Double Duplex	Includes base, receptacle, cover, & misc. hardware. 15 amp. receptacles 20 amp. receptacles	80.75 83.75
Combination Power/Communications	Includes base, receptacle, cover, & misc. hardware. 15 amp. receptacle only + connector for 1 phone + connector for 2 phones 20 amp. receptacle only + connector for 1 phone + connector for 2 phones	66.75 89.75 94.00 70.00 93.00 97.00
Cable Fold (For a change in direction)	No materials required	5.00

F

F-32

F-33

COST INSTALLED IN DOLLARS PER FOOT

$25

20

15

10
9
8

7

6

5

Blank Insert — Single
Blank Insert — Double
Blank Insert — Triple

**Standard Duct
1 3/8 x 3 1/8**

Blank Insert — Single
Blank Insert — Double

**Super Duct
1 3/8 x 7 1/4**

Blank Insert — 1-Std 1-Super
Blank Insert — 2-Std 1-Super
Blank Insert — 1-Std 2-Super

The costs shown for the underfloor duct consist of the published contractors' book prices for the ducts, couplings, duct supports every 5 feet, marker screws for all inserts, sealing compound, and the labor for installation in a 4- to 6-inch concrete floor. It is assumed that one manual cut must be made for each 100 feet of duct. No wiring costs are included. The duct is manufactured by Square D Co.

COST OF UNDERFLOOR DUCT JUNCTION BOXES AND FITTINGS

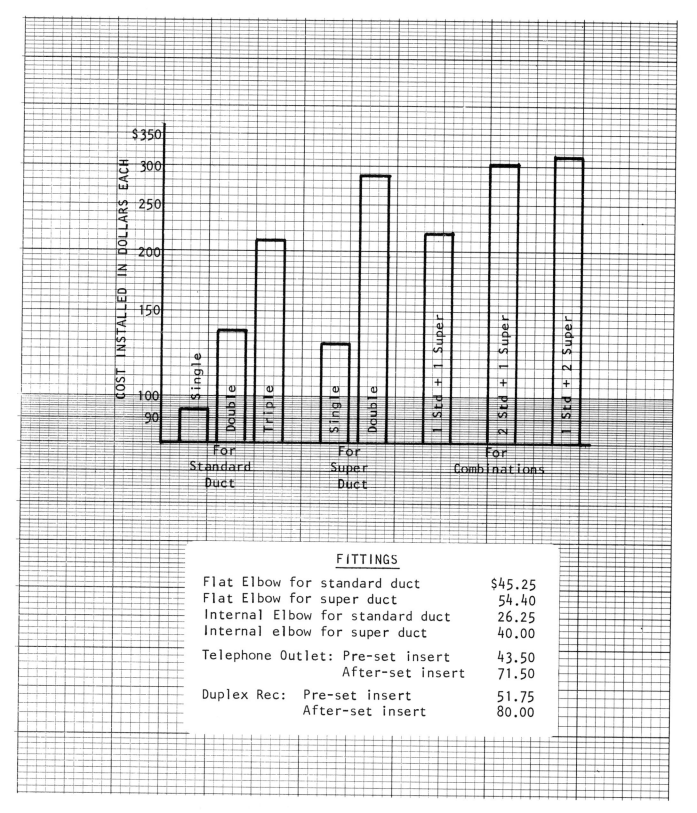

COST INSTALLED IN DOLLARS EACH

- Single — For Standard Duct
- Double
- Triple
- Single — For Super Duct
- Double
- 1 Std + 1 Super — For Combinations
- 2 Std + 1 Super
- 1 Std + 2 Super

FITTINGS

Flat Elbow for standard duct	$45.25
Flat Elbow for super duct	54.40
Internal Elbow for standard duct	26.25
Internal elbow for super duct	40.00
Telephone Outlet: Pre-set insert	43.50
After-set insert	71.50
Duplex Rec: Pre-set insert	51.75
After-set insert	80.00

F-34

The costs shown for the single-level underfloor duct junction boxes and fittings consist of the published contractors' book prices for the boxes and fittings shown. They also include the labor for installation in a 4- to 6-inch concrete floor.

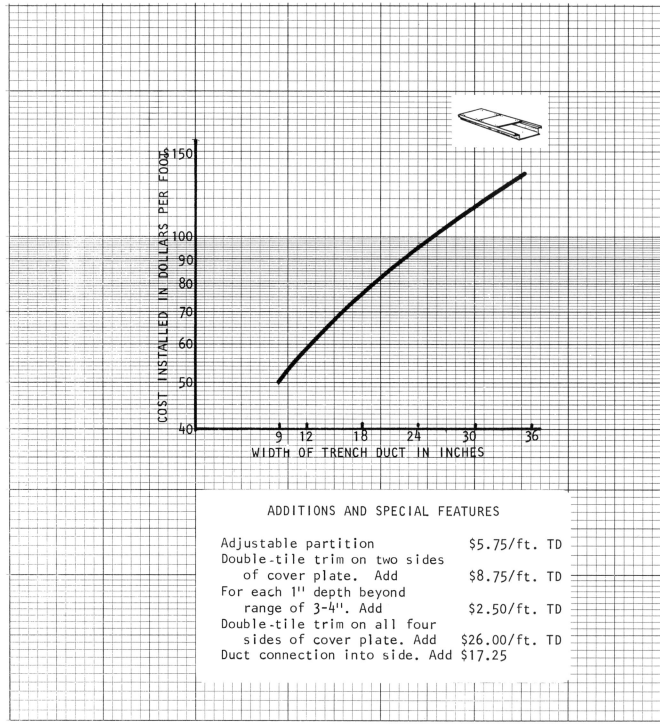

ADDITIONS AND SPECIAL FEATURES

Adjustable partition	$5.75/ft. TD
Double-tile trim on two sides of cover plate. Add	$8.75/ft. TD
For each 1" depth beyond range of 3-4". Add	$2.50/ft. TD
Double-tile trim on all four sides of cover plate. Add	$26.00/ft. TD
Duct connection into side. Add $17.25	

The costs shown for the trench duct consist of the contractors' published book prices for the sizes shown. Labor for installation is included but freight is not included. The trench duct shown here is as manufactured by Square D Co. It is considered to be tack-welded to a cellular floor system. The standard straight lengths are 10 feet long and include five cover plates 2 feet long and 1/4 inch thick, of painted steel, for each 10-foot length of duct. The nominal trench width is actually the cover-plate width, and the actual inside trench width is about 2 inches less. The standard duct is furnished with single-tile trim for 1/8-inch floor tile. The trench is single-compartment with an adjustable depth (before pouring concrete) from 2 3/8 to 3 3/8 inches.

F-35

TRENCH DUCT & COVERS

The costs shown for trench duct fittings consist of the contractors'
published book prices for the sizes indicated and the labor for installation.
These are all companion fittings to the duct and are essentially self-
explanatory. The vertical elbow, riser, and cabinet connector are for
connecting the duct system to a telephone, signal, or a power panel.

COST OF TRENCH DUCT FITTINGS

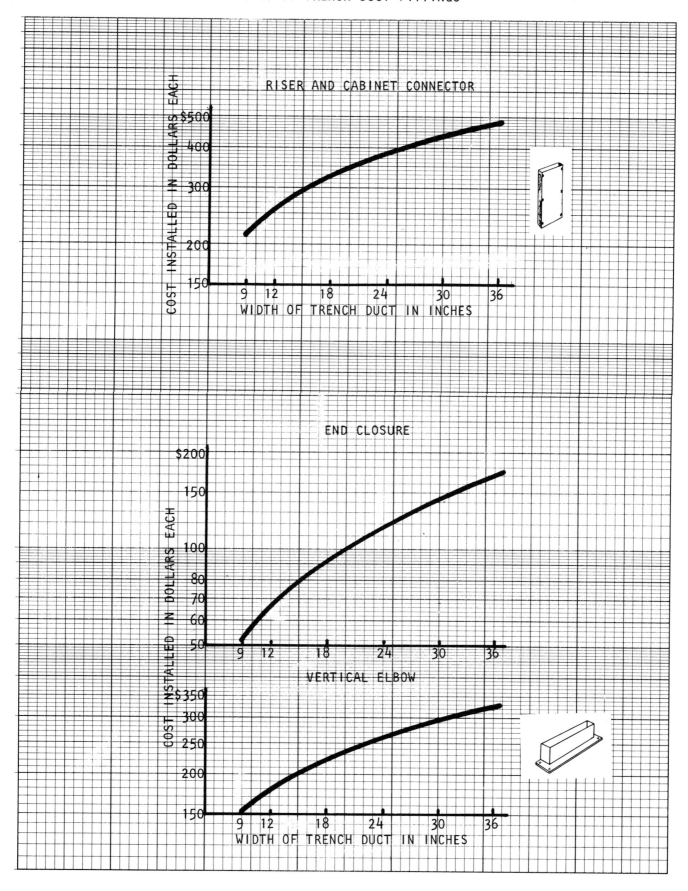

RISER AND CABINET CONNECTOR

COST INSTALLED IN DOLLARS EACH

$500
400
300
200
150

9 12 18 24 30 36

WIDTH OF TRENCH DUCT IN INCHES

END CLOSURE

COST INSTALLED IN DOLLARS EACH

$200
150
100
80
70
60
50

9 12 18 24 30 36

VERTICAL ELBOW

$350
300
250
200
150

9 12 18 24 30 36

WIDTH OF TRENCH DUCT IN INCHES

*The costs shown for the trench duct fittings consist of the contractors'
published book prices for the sizes indicated and the labor for installation.
These are all companion fittings to the duct and are self-explanatory.*

COST OF TRENCH DUCT FITTINGS

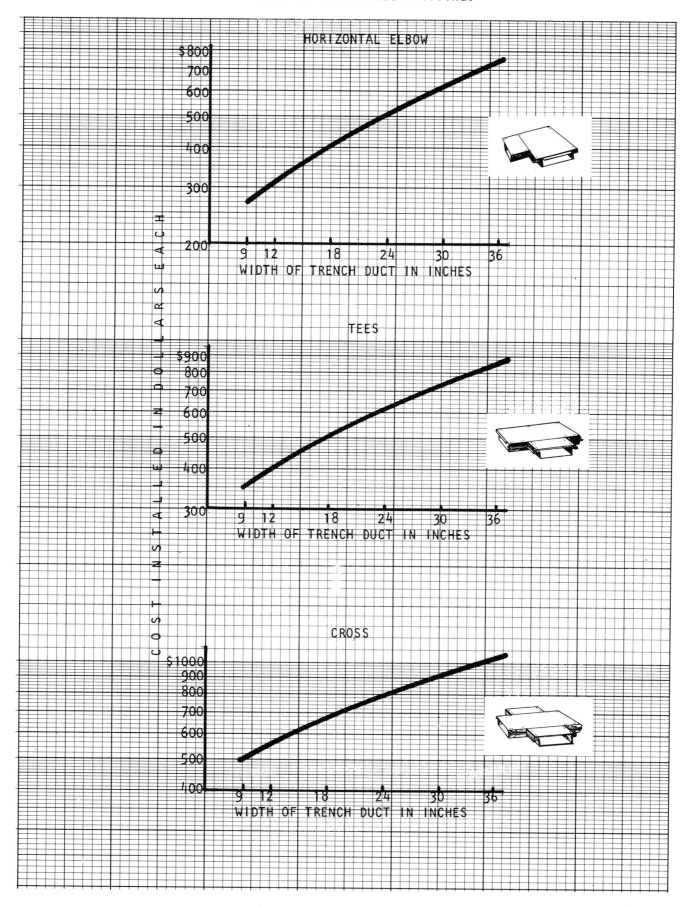

HORIZONTAL ELBOW

WIDTH OF TRENCH DUCT IN INCHES

TEES

WIDTH OF TRENCH DUCT IN INCHES

CROSS

WIDTH OF TRENCH DUCT IN INCHES

COST INSTALLED IN DOLLARS EACH

The costs shown for these units consist of the contractors' published book prices for the units shown. This lighting duct system is manufactured by I-T-E and known as universal lighting duct, in the sizes and capacities shown. Installation is assumed to be made on an existing ceiling.

Included: (material and labor)
   Hanger supports for surface mounting every 5 feet
   Toggle bolts for fastening hangers
   Plain coupling every 10 feet

COST OF LIGHTING DUCT

DESCRIPTION	Installed Cost			
	250 Volt - 2 Wire		125/250 Volt	300 Volt
	20-25 Amp	50-60 Amp	50-60 Amp 3 Wire	50 Amp 2 Wire
Duct Section	$5.25/Ft.	$5.75/Ft.	$7.00/Ft.	$6.75/Ft.
End Cap	$10.50	$10.50	$10.50	$10.50
End Feed-In Box	$57.25	-----	$63.50	$61.50
Center Feed-In Box	$69.00	-----	$76.00	$68.25

The twist-out terminal plug is designed for accepting a two-wire portable cord with grounding conductor. No wire is included in this price.

The twist-out receptacle will accept a two-wire plug with grounding prong.

The trolley terminal and receptacle are similiar to the twist-out except that they are free to roll along the duct for maximum flexibility.

Feed-in boxes include 3/4-inch EMT terminal and toggle bolts for mounting.

COST OF LIGHTING DUCT

DESCRIPTION	Installed Cost			
	250 Volt - 2 wire		125/250 Volt	300 Volt
	20 Amp	50 Amp	50 Amp - 3Wire	50 Amp - 2Wire
Couplings - Standard	$8.00	$8.00	$10.75	$9.00
Trolley Entrance	$17.00	$17.00	-----	$17.00
Twist-Out Terminal Plug	$13.50	$13.50	$14.25	$14.25
Twist-Out Receptacle	$12.50	$12.50	$12.50	-----
Trolley Terminal	$21.25	$21.25	-----	-----
Trolley Terminal - Heavy Duty	$35.50	$35.50	-----	-----

FITTINGS FOR LIGHTING DUCT

Control Equipment

LIGHTING CONTROL

Low-Voltage Control System

Switches	G-1
Control System Components	G-2
Cables for Low-Voltage System	G-3

Photoelectric Control Switches — G-4

Time Switches

24-Hour & Astronomic Dial	G-5

Multipole Relays

Mechanically and Electrically Held	G-6
Silent Mercury—Electrically Held	G-7

Lighting Contactors

Electrically Held	G-8
Mechanically Held	G-9

DIMMERS

Incandescent Wall Dimmers	G-10
Incandescent Manual Remote	G-11
Incandescent Motorized Remote	G-12
Fluorescent Wall Dimmers	G-13
Fluorescent Manual Remote	G-14
Fluorescent Motorized Remote	G-15

POWER CONTROL

Individual Motor Starters—Line Voltage

Small-Motor Hookups and Control Switches	G-16
Line Voltage—200, 230, and 460 Volt—Three-Phase	G-17
Combination Motor Starter—Switch & Fuse Type	G-18

Individual Reduced-Voltage Starters

Autotransformer—Closed Transition	G-19
Wye-Delta—Open Transition	G-20
Part-Winding—Two-Step—Closed Transition	G-21
Primary Resistance Type—Nonreversing	G-22

G

Power Factor Corrective Capacitors   G-23

MOTOR CONTROL CENTERS

*Vertical Section—20 Inch Deep—with MLO,
  Main Switch or Breaker*   G-24
*Vertical Section—20 Inch Deep—for Starters
  or Disconnects*   G-25
Fusible Switch Disconnects   G-26
Circuit-Breaker Disconnects   G-27
Full-Voltage Starter—Single-Speed—Nonreversing   G-28
Full-Voltage Starter—Single-Speed—Reversing   G-29
Full-Voltage Starter—Two-Speed—Nonreversing—Two Winding   G-30
Reduced Voltage—Autotransformer—Closed Transition   G-31
Reduced Voltage—Part Winding—Two Step   G-32

The costs shown for the low-voltage control system components consist of the published contractors' book prices and the labor for installation. All equipment is now specification-grade.

Local switches are considered to be installed flush in a masonry wall in either a single- or a two-gang box with a deep tile ring, two 1/2-inch EMT terminals, the necessary low-voltage switches, and the appropriate wall plate.

The selector type of switch, 8- or 12-position, is also considered to be installed flush in a masonry wall in a two-gang box with a deep tile ring, two 1/2-inch EMT terminals, the necessary switch, and the appropriate wall plate.

The motorized master switch is considered to be installed in an RB-3 component cabinet and flush in a masonry wall. The labor applied allows for hooking up all wire terminals and installing the enclosure and switch.

The component cabinets for the relays are considered to be installed flush in a masonry wall.

All equipment is manufactured by General Electric Co.

COST OF LOW-VOLTAGE REMOTE CONTROL SYSTEM

LOCAL CONTROL SWITCH	Type of Plate	COST INSTALLED IN DOLLARS EACH			
		Normal Switch		Key Operated Switch	
		No Pilot Light	With Pilot Light	No Pilot Light	With Pilot Light
1 Position, 1 Gang	Nylon	$40.00	$43.25	$44.50	$48.25
	Stainless	41.50	44.75	46.00	49.50
2 Position, 1 Gang	Nylon	48.00	54.75	57.25	64.50
	Stainless	49.50	56.25	58.50	66.00
3 Position, 2 Gang	Nylon	58.50	68.50	72.00	83.25
	Stainless	62.25	72.25	75.75	87.00
4 Position, 2 Gang	Nylon	67.50	81.00	84.75	100.50
	Stainless	71.00	84.25	89.00	103.75
8 Position, 2 Gang Master Switch	Anodized Aluminum	118.75	145.50	155.00	184.50

COST OF LOW-VOLTAGE REMOTE CONTROL SYSTEM

Description	Installed Cost	Description	Installed Cost
Wall Switch: Conventional Momentary ON-OFF-ON Locking Type	$42.75 46.75	Silicon Rectifier	$27.00
Wall Switch: Interchangeable	42.00	Blocking Diode Assembly	$151.00
Master Control Switch: 2 dial		Component Cabinets	
RMS-4A	$122.25	RBS-1 8" x 12"+cover	68.00
With Pilot Light Assy.	211.50	RBS-2 12" x 12" + cover	82.75
Motorized Master Control		RB-3 18" x 28" + cover	396.25
'ON' Unit	$144.75		
'OFF' Unit	144.75	Frames	
Relays		RRF-78 Holds 1-6 relays	34.25
RR-7 No P.L.Sw.	24.25	RFT-178 With 118/24 v xfmr.& rect. Will hold 3 relays.	113.50
RR-8 With P.L.Sw.	27.25		
RR-9 Isolated P.L.Sw.	30.50	RFT-278 With 277/24 v xfmr.& rect. Will hold 3 relays	134.25
Transformer		Remote Control Interface	
RT-1 120/24v	69.50		
RT-2 277/24v	90.00	RCI-1	141.50

G-2

LOW-VOLTAGE REMOTE CONTROL SYSTEM COMPONENTS

COST OF LOW-VOLTAGE REMOTE CONTROL SYSTEM

Description		Installed Cost
Multiple Relay Energizer - RCE-1		
Makes possible the switching of up to 12 relays from up to 2000' with #20 wire.		$141.50
Twisted (No jacket) Indoor Wire		
	2 - #20	.39/ft.
	3 - #20	.44/ft.
	4 - #20	.51/ft.
Multi-conductor (Jacketed) Indoor Wire		
	12 - #20	1.57/ft.
	19 - #20	1.85/ft.
	26 - #20	2.60/ft.
Flat Outdoor Wire		
	3 - #18	.69/ft.

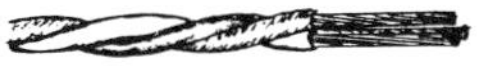

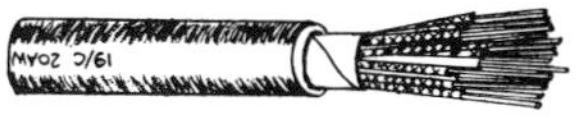

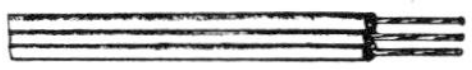

The costs shown for the switches consist of the published contractors' book price and the labor for installation onto the surface of a masonry wall.

The timer switch is as manufactured by N. H. Rhodes, Inc., or Intermatic Incorporated. The other time and photoelectric switches are manufactured by Tork.

COST OF PHOTOELECTRIC & TIMER SWITCHES

DESCRIPTION		Installed Cost

7 Day Dial	Standard:
	    Indoor: SPST, DPST, 3PST, DPDT
	        Momentary Type: SPDT
	
	Reserve Power:
	    Indoor/Outdoor: SPST, DPST, 3PST, DPDT
	    Momentary Type: SPDT
		$144.25
		298.00
		
		294.50
		451.50

Photo/Time	Reserve Power & Hand-Off-Auto Switches
	
	   2 Circuit - Dusk/Time, Dusk/Dawn SPDT
	
	   3 Circuit - Dusk/Time, Dusk/Dawn,
	             Time/Time, SPDT
	
	   These include a photoelectric control
		652.25
		
		693.25

Turn-lock Type	Photoelectric Control
	
	   1800 watt, 120 volt. Includes
	                    mounting bracket
		56.25

Conduit Type	Photoelectric Control 2000 watt, 120 volt	53.25
Flush Mounting	Photoelectric Control 1800 watts, 120 volt	72.00

Contactor Type	Photoelectric Control
	
	   6000 watt, 120 volt, DPST, with Turn-lock
	   photoelectric control and raintight
	   contactor enclosure.
		168.50

	Flush mounted wall box timer switch	57.00

DESCRIPTION		COST INSTALLED EACH
24 Hour Dial	Standard Units - 40 Amp. On/Off	
	Indoor - SPST	$103.25
	Indoor - DPST	108.25
	Outdoor SPST	124.00
	Outdoor DPST	129.00
	Program Time Switch - SPDT	136.25
	Standard Units with Skip-A-Day feature	
	Indoor - SPST	115.50
	Indoor - DPST & SPDT	123.00
	Indoor - 3PST & DPDT	149.50
	Standard Units with Skip-A-Day & Reserve Power	
	Indoor - DPST & SPDT	272.50
	Indoor - 3PST & DPDT	301.25
Astronomic Dial MODEL 7200Z	Skip-A-Day feature	
	Indoor - DPST & SPDT	139.25
	Indoor - 3PST	176.25
	Skip-A-Day and Reserve Power	
	Indoor - DPST & SPDT	283.25
	Indoor - 3PST	318.25

The costs shown for these time switches consist of the published contractors' book price with anchors and the labor for installing the switches onto the surface of a masonry wall. The switches shown are manufactured by Tork.

G-5

COST OF 20-AMP MULTIPOLE LIGHTING CONTACTORS

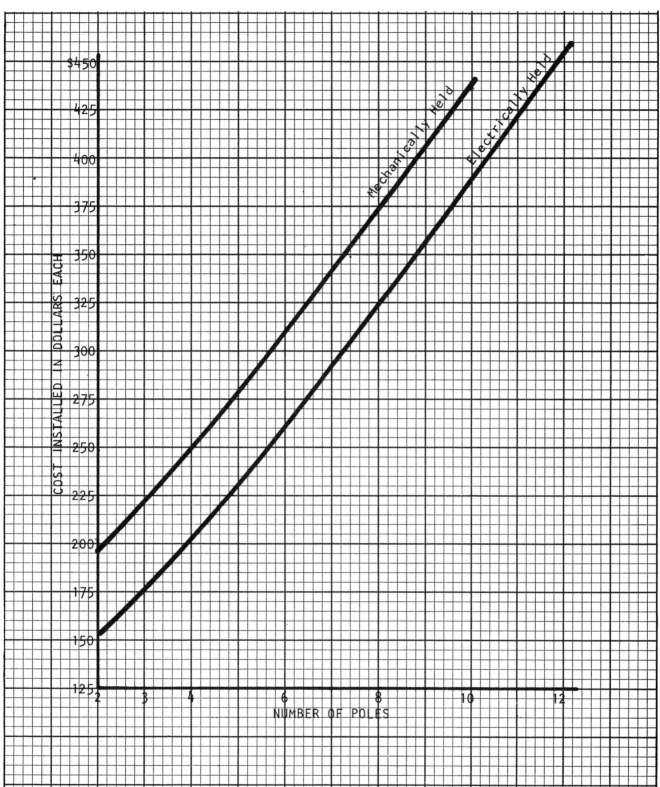

The light-duty lighting contactors or relays of the multipole type are a
Square D 8903 Series with 20-ampere poles mounted in a general-
purpose enclosure and fastened to a masonry wall. Included are conduit
terminals and wire within the enclosure. No control switch for either
the electrically held or the mechanically held unit is included. The units
are rated for 480-volt-maximum line voltage, and the contacts are such
that you may apply full rating for tungsten, fluorescent, or mercury
vapor lighting fixtures.

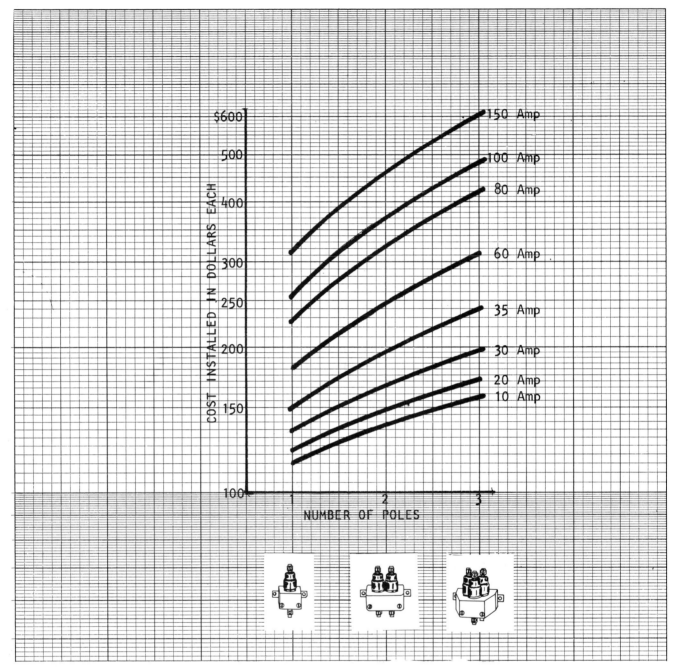

The costs shown for these relays consist of the published contractors' book prices for the size and number of poles shown. These relays are manufactured by Durakool, Inc., 1080 North Main Street, Elkhart, Ind. 46514, and are appropriate for applications requiring quiet relay operation.

Included: (material and labor)

Screw-cover junction box of a size to accept a relay with fastening devices for surface mounting to a masonry wall.

EMT terminals to suit conduit size required.

Sufficient wire of size and strength to suit the requirements inside the unit and for making up connections.

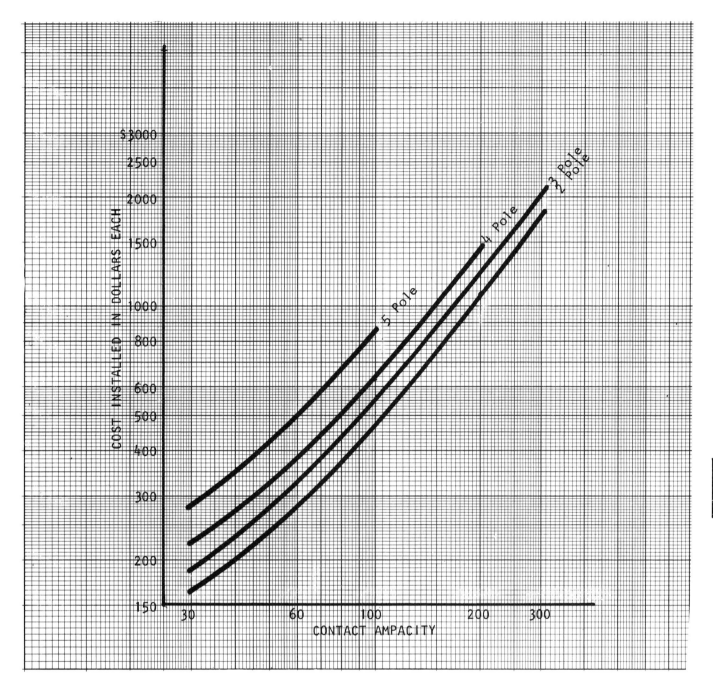

The costs shown for these contactors consist of the contractors' book prices for the items shown. The units are rated for 480-volt-maximum line voltage. These units are manufactured by Square D Co. and are of Class 8903 in a general-purpose enclosure. You may apply full rating for tungsten, fluorescent, and mercury vapor lighting fixtures.

Included: (material and labor)

Fastening devices and mounting to a masonry wall.

EMT terminals.

Sufficient wire of size and length to suit the requirements inside the unit.

Making connections.

G

G-8

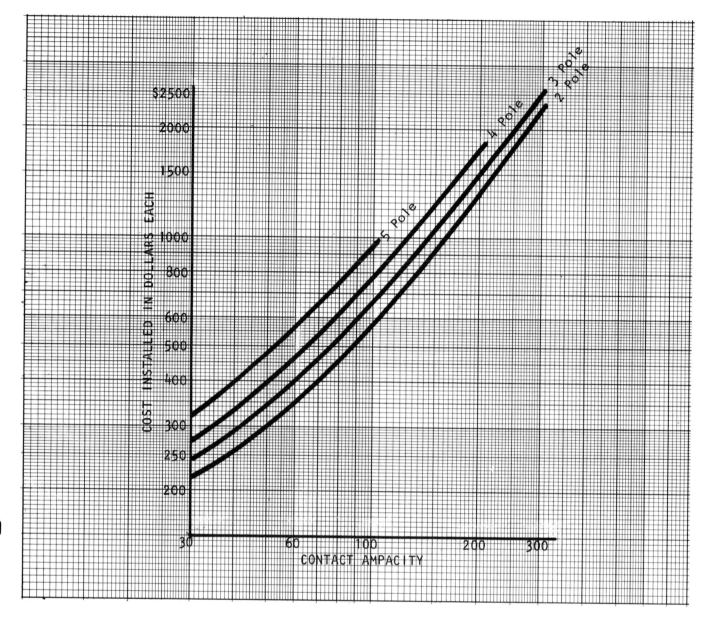

The costs shown for these contactors consist of the contractors' book prices for the items shown. The units are rated for 480-volt-maximum line voltage. These units are manufactured by Square D Co. and are of Class 8903 in a general-purpose enclosure. You may apply full rating for tungsten, fluorescent, and mercury vapor lighting fixtures.

Included: (material and labor)
  Fastening devices and mounting to a masonry wall.

  EMT terminals.

  Sufficient wire of size and length to suit the requirements inside the unit.

  Making connections.

G-9

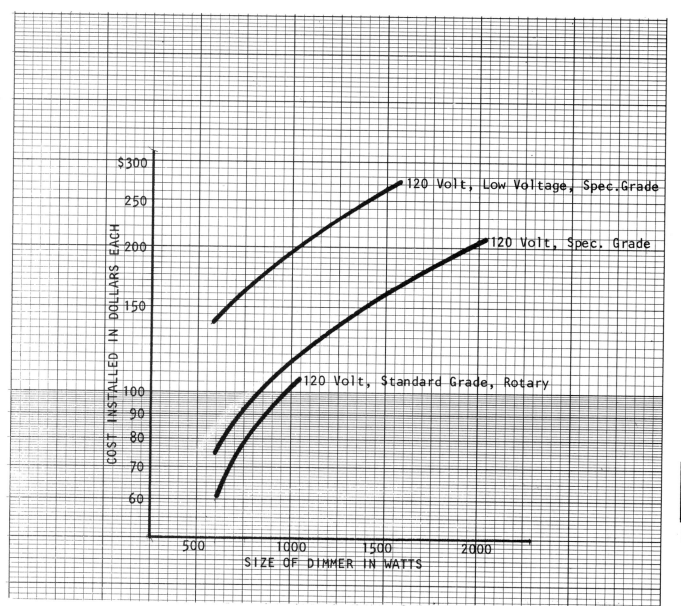

The costs shown for the incandescent wall dimmers consist of the published contractors' book prices for the units shown. The costs also include a flush wall box with two conduit terminals. These dimmers are manufactured by Lutron Electronics Co., Inc., and the 120-volt specification grade is the Nova series with a slider-type control with an ON-OFF switch. The standard unit is the low-cost type with a rotary control with a push-type ON-OFF switch.

G

G-10

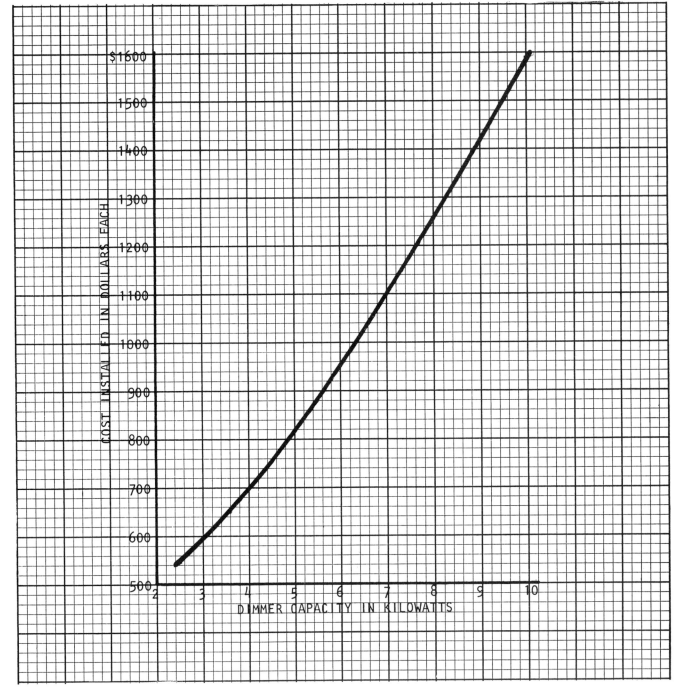

G-11

The cost of these dimmers includes the published contractors' book prices and the labor for installing the units on the wall and connecting the wires in the units. No interconnecting raceway is included.

These single-phase manual remote incandescent dimmers are made up of modules including the lighting control station, master control card, and lighting controllers. The lighting control station looks like a low-powered wall dimmer. These units are manufactured by Lutron Electronics Co., Inc.

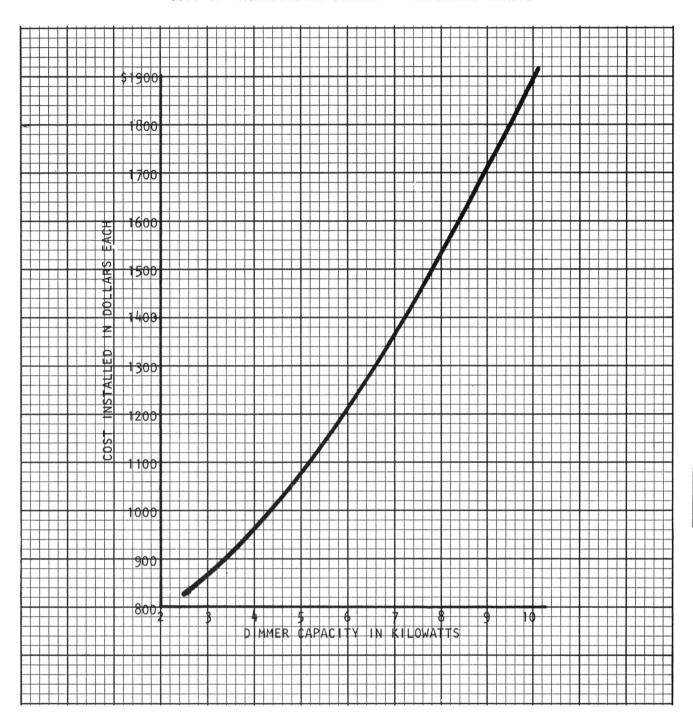

The cost of these dimmers includes the published contractors' book prices and the labor for installing the units on the wall and connecting the wires in the units. No interconnecting raceway or wire is included.

These single-phase motorized remote incandescent dimmers consist of modules including the lighting control station, master control card, lighting controllers, and a multistation control interface which contains the motor drive unit. The lighting control station looks like a low-powered wall dimmer. These units are manufactured by Lutron Electronics Co., Inc.

G

G-12

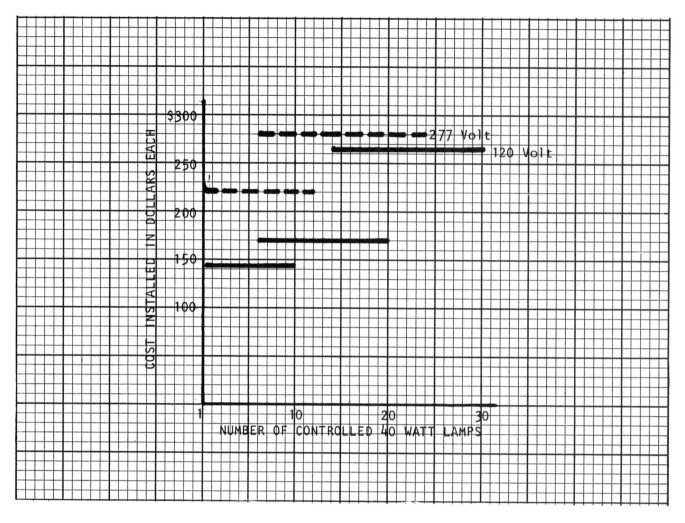

G-13

The costs shown for the fluorescent wall dimmers consist of the published contractors' book prices for the units shown and the associated labor. The costs also include a flush wall box with two conduit terminals.

These dimmers are Nova series manufactured by Lutron Electronics Co., Inc., and have a slider-type control with ON-OFF switch.

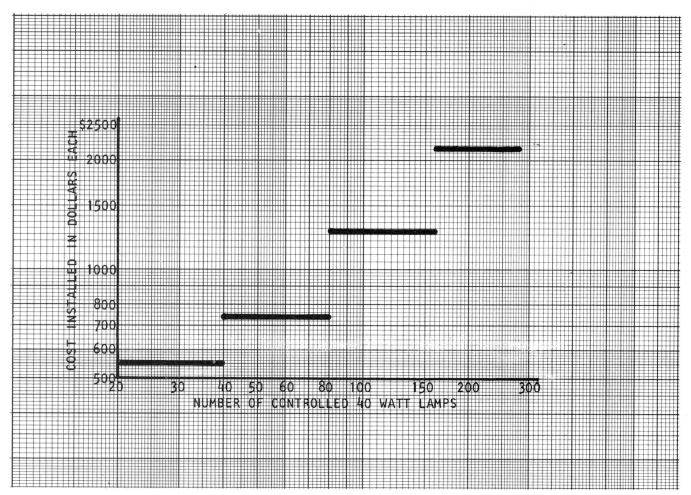

The costs shown for these fluorescent dimmers include the published contractors' book prices and the labor for installing the units on a wall and connecting the wires in the units. No interconnecting raceway or wire is included.

These single-phase manual remote fluorescent dimmers are made up of modules consisting of a lighting control station, lighting controller, and lighting control card. The lighting control station looks like a low-powered wall dimmer. These units are manufactured by Lutron Electornics Co., Inc.

G

G-14

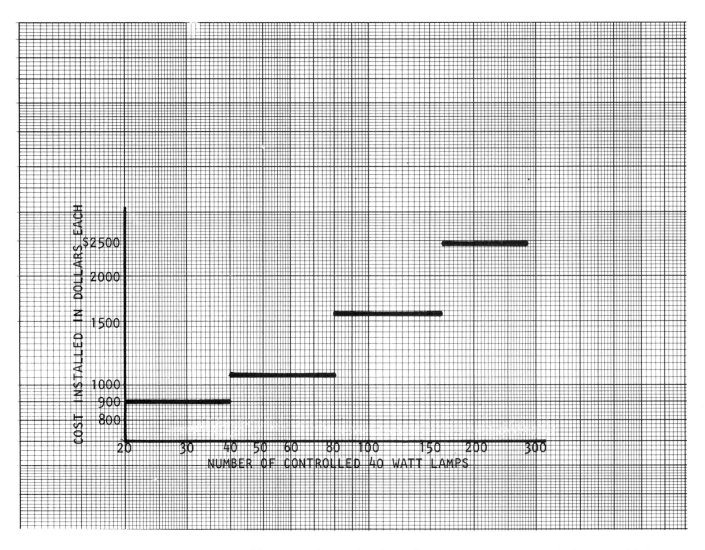

G-15

The costs shown for these fluorescent dimmers include the published contractors' book prices and the labor for installing the units on a wall and connecting the wires in the units. No interconnecting raceway or wire is included.

These single-phase motorized remote dimmers consist of modules including a lighting control station, a master control card, lighting controllers, and a multistation control interface which contains the motor drive unit. The lighting control station looks like a low-powered wall dimmer. These units are manufactured by Lutron Electronics Co., Inc.

HAND MOTOR STARTERS AND HOOKUPS

	115 Volt – 1 Pole		208 or 230 Volt 2P		208 or 230 Volt 3P	
	General Purpose	Weatherproof	General Purpose	Weatherproof	General Purpose	Weatherproof
1/6 – 1.0 H.P.	$73.75	$122.00	$78.25	$126.25	NA	NA
1½ – 3.0 H.P.	NA	NA	137.00	264.00	$149.75	$257.75
5 – 7½ H.P.	NA	NA	NA	NA	165.75	283.50

CONTROL SWITCHES

Description		Standard Duty	Heavy Duty
Pushbutton Type	Single unit	$45.25	$104.75
	Two unit	57.00	118.25
	Three unit	85.00	178.00
	Two unit with pilot light	112.25	191.75
Selector Switch Type	Two position	45.25	107.50
	Three position	59.00	114.00

The costs shown for these hand motor starters and hookups consist of the published contractors' book prices for the switches required and the labor for the installation. The prices include the heating elements, a 3-foot length of flexible steel conduit, wire, conduit, and wire connectors. These units are the Allen-Bradley 600 and 609 Series.

G

G-16

The costs shown for the motor starters consist of the published contractors' book price for the motor starters required for the horsepower and voltages shown. They further include fastening devices for securing to a masonry wall, a 3-foot length of flexible metal conduit with 4 feet of conductor, the necessary fittings and wire terminals, and the labor for making a complete installation. When the starter is furnished by others, it is considered that only the starter, with the appropriate heaters, is furnished to the electrical contractor.

NEMA Size	LINE VOLTAGE		
	HORSEPOWER RANGE		
	200-Volt Motor	230-Volt Motor	460-Volt Motor
00	0-1 1/2	0-1 1/2	0-2
0	2-3	2-3	3-5
1	4-7 1/2	4-7 1/2	6-10
2	8-10	8-15	11-25
3	11-25	16-30	26-50
4	26-40	31-50	51-100
5	41-75	51-100	101-200
6	76-150	101-200	201-400

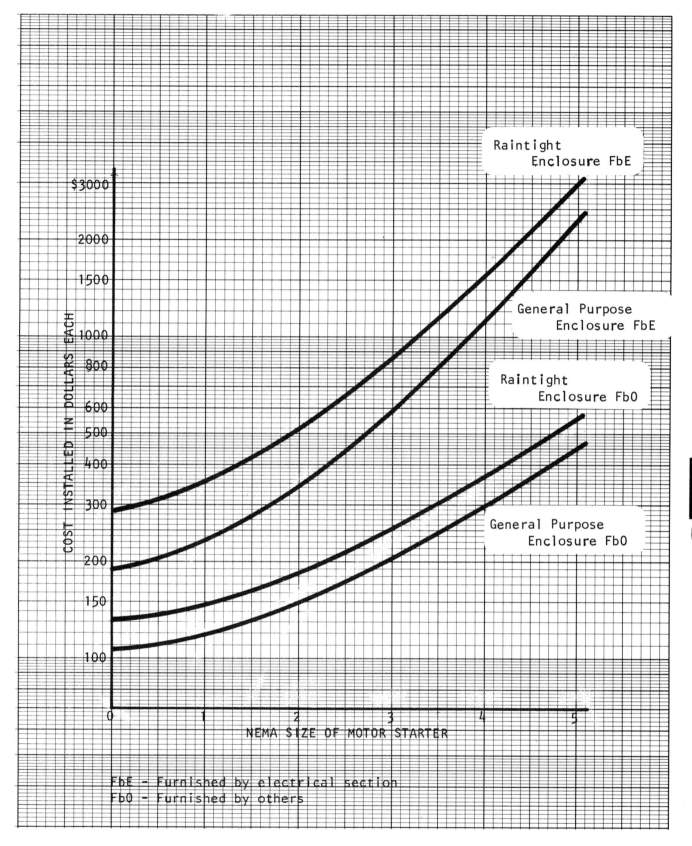

Raintight
Enclosure FbE

General Purpose
Enclosure FbE

Raintight
Enclosure FbO

General Purpose
Enclosure FbO

FbE - Furnished by electrical section
FbO - Furnished by others

COST INSTALLED IN DOLLARS EACH

NEMA SIZE OF MOTOR STARTER

G

G-17

200, 230, and 460 VOLT AC LINE VOLTAGE MAGNETIC STARTER - 3 PHASE

The costs shown for the combination motor starters consist of the contractors' published book prices for the motor starters required for the horsepower and voltages shown. The starters are Square D 8538 Series. They include fastening devices for securing to a masonry wall, a 3-foot length of flexible metal conduit with 4 feet of conductor at the motor, the necessary fittings and wire terminals, and the labor for making a complete installation with the exception of the length of conduit and wire run from the starter to the motor. Weatherproof installations provide liquidtight flexible conduit.

COMBINATION MOTOR STARTER			
NEMA Size	HORSEPOWER RANGE		
	200-Volt Motor	230-Volt Motor	460-Volt Motor
0	0-3	0-3	0-5
1	4-7 1/2	4-7 1/2	6-10
2	8-10	8-15	11-25
3	11-25	16-30	26-50
4	26-40	31-50	51-100
5	41-75	51-100	101-100
6	76-100	101-200	201-400

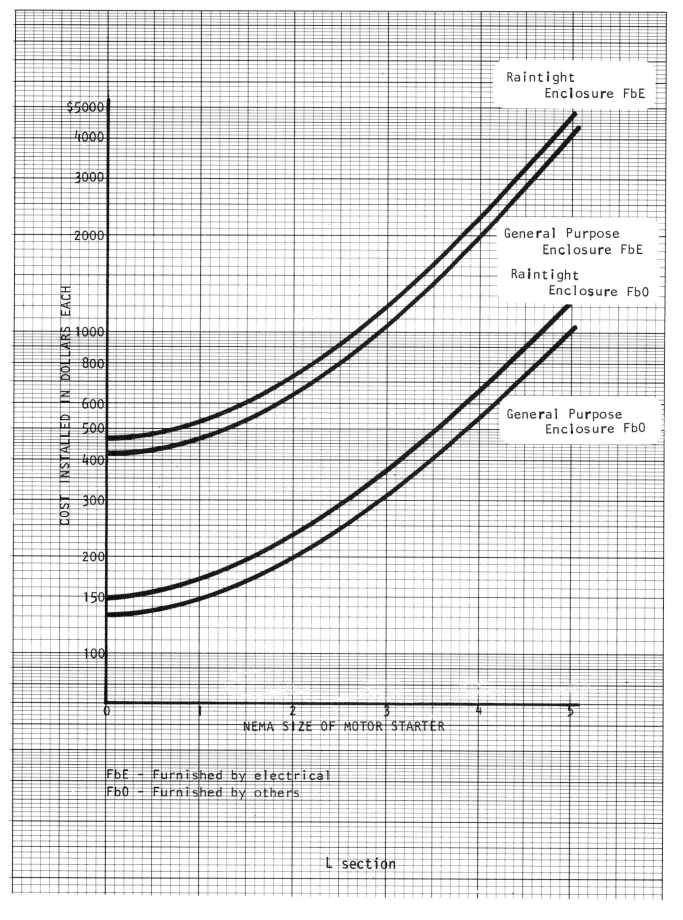

Raintight
Enclosure FbE

General Purpose
Enclosure FbE

Raintight
Enclosure Fb0

General Purpose
Enclosure Fb0

COST INSTALLED IN DOLLARS EACH

NEMA SIZE OF MOTOR STARTER

FbE - Furnished by electrical
Fb0 - Furnished by others

L section

G-18

The costs shown for the reduced-voltage starters are based upon the published contractors' book prices. These are Square D Class 8606.

Included: (material and labor)

Start/stop pushbutton in cover.

Pilot light.

Toggle bolts for mounting to masonry wall.

Conduit terminals.

Sufficient conductor of proper size and length required in the starter enclosure for connections.

3 feet of flexible conduit, terminals, 4 feet of the number and size of conductors required, and connections.

3 overload thermal units.

AUTOTRANSFORMER			
NEMA Size	HORSEPOWER RANGE		
	200-Volt Motor	230-Volt Motor	460-Volt Motor
2	0-10	0-15	0-25
3	11-25	16--30	26-50
4	26-40	31-50	51-100
5	41-75	51-100	101-200
6	76-150	101-200	201-400

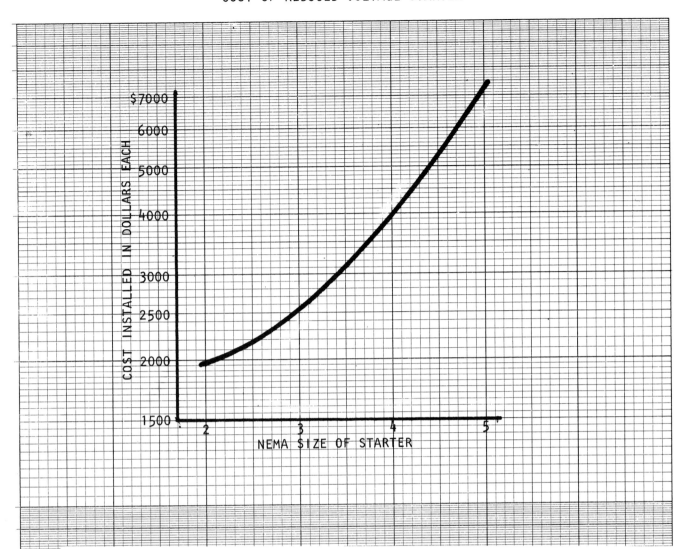

COST INSTALLED IN DOLLARS EACH

NEMA SIZE OF STARTER

G

G-19

REDUCED-VOLTAGE STARTER - AUTOTRANSFORMER - CLOSED TRANSITION

The costs shown for the reduced-voltage starters are based upon the published contractors' book prices. These are Square D Class 8630.

Included: (material and labor)

Start/stop pushbutton in cover.
Pilot light.

Toggle bolts for mounting to masonry wall.

Conduit terminals.

Sufficient conductor of proper size and length required in the starter enclosure for connections.

3 feet of flexible conduit, terminals, 4 feet of the number and size of conductors required, and connections.

3 overload thermal units.

WYE DELTA			
NEMA Size	HORSEPOWER RANGE		
	200-Volt Motor	230-Volt Motor	460/575-Volt Motor
1YD	0-10	0-10	0-15
2YD	11-20	11-25	15-40
3YD	21-40	26-50	41-75
4YD	41-60	51-75	76-100
5YD	61-150	76-150	151-300

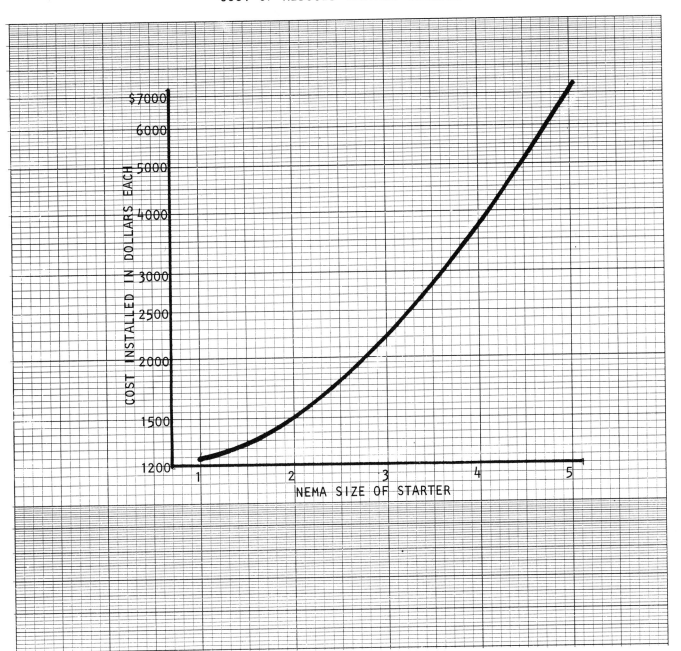

REDUCED-VOLTAGE STARTER - WYE-DELTA TYPE - OPEN TRANSITION

The costs shown for the reduced-voltage starters are based upon the published contractors' book prices. These are Square D Class 8640.

Included: (material and labor)

Start/stop pushbutton in cover.

Pilot light.

Toggle bolts for mounting to masonry wall.

Conduit terminals.

Sufficient conductor of proper size and length required in the starter enclosure for connections.

3 feet of flexible conduit, terminals, 4 feet of the number and size of conductors required, and connections.

3 overload thermal units.

NEMA Size	PART WINDING		
	HORSEPOWER RANGE		
	200-Volt Motor	230-Volt Motor	460-Volt Motor
1PW	0-10	0-10	0-15
2PW	11-20	11-25	16-40
3PW	21-40	26-50	41-75
4PW	41-75	51-75	76-150
5PW	76-150	76-150	151-350
6PW	-------	151-300	351-600

COST OF REDUCED-VOLTAGE STARTER

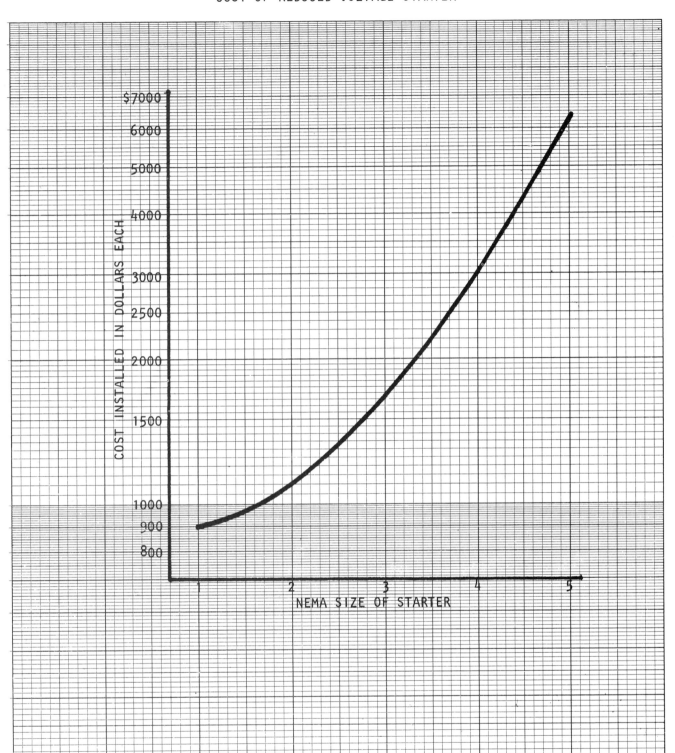

REDUCED-VOLTAGE STARTER - PART-WINDING TYPE - CLOSED TRANSITION

The costs shown for the reduced-voltage starters are based upon the published contractors' book prices. These are Square D Class 8547.

Included: (material and labor)

Start/stop pushbutton in cover.

Pilot light.

Toggle bolts for mounting to masonry wall.

Conduit terminals.

Sufficient conductor of proper size and length required in the starter enclosure for connections.

3 feet of flexible conduit, terminals, 4 feet of the number and size of conductors required, and connections.

3 overload thermal units.

PRIMARY-RESISTOR TYPE			
NEMA Size	HORSEPOWER RANGE		
	200-Volt Motor	230-Volt Motor	460-Volt Motor
1	0-7 1/2	0-7 1/2	0-10
2	8-10	8-15	11-25
3	11-25	16-30	26-50
4	26-40	31-50	51-100
5	41-75	51-100	101-200

COST OF REDUCED-VOLTAGE STARTER

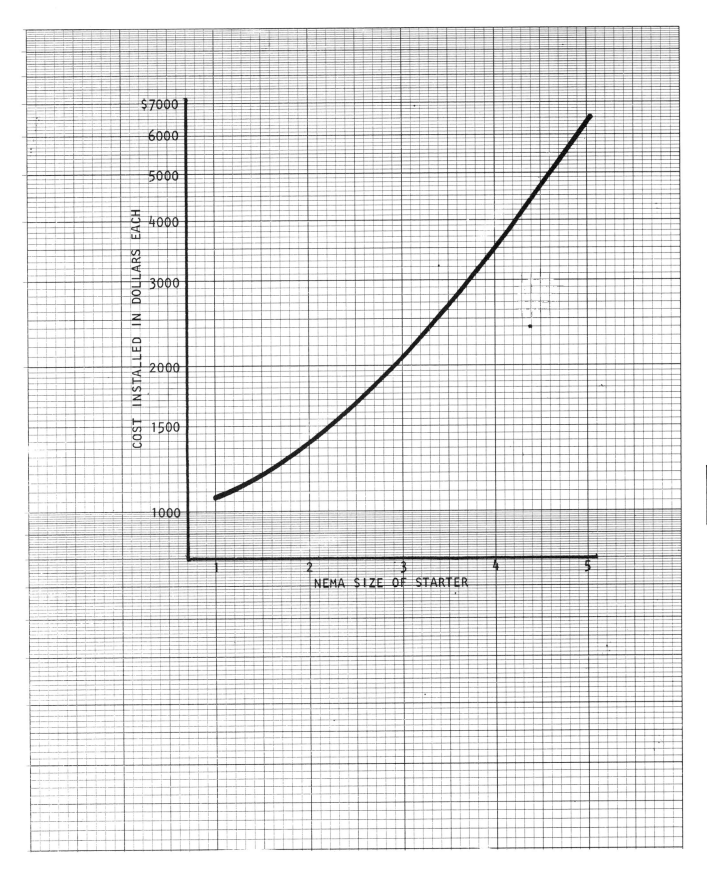

REDUCED-VOLTAGE STARTER - PRIMARY-RESISTOR TYPE - NONREVERSING

The costs shown for the power factor corrective capacitors consist of the published contractors' book prices as manufactured by Sprague Electric Company. The capacitors are internally fused. Included are mounting brackets, conduit terminals, and wire terminations for #6 and larger. Labor is included.

COST OF POWER FACTOR CORRECTIVE CAPACITORS

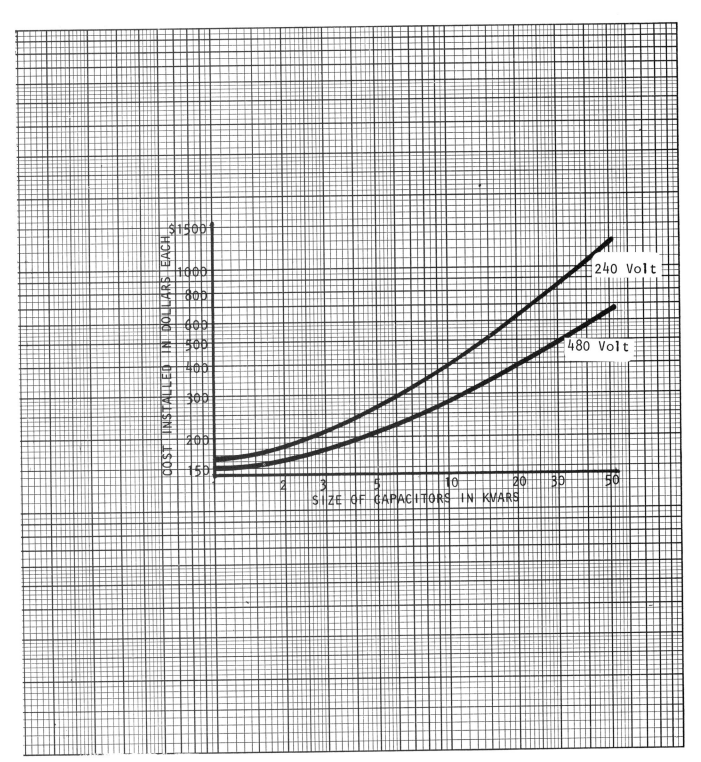

POWER FACTOR CORRECTIVE CAPACITORS - THREE-PHASE - 240 AND 480 VOLT

The control center components used in establishing these graphs are based upon Square D Model 4 motor control centers.

The installed cost of a free-standing type of motor control center can be assembled from the use of a few pages of the essential components as

1. Vertical sections required.
2. Outgoing feeder disconnects.
3. Motor starters as required.

Prices are based upon Class I, type B, wiring, which provides factory-wired terminal boards for load and control wires in the starter units. This seems to be the most popular type.

VERTICAL SECTION FOR INCOMING FEEDER

The vertical section is a 20-inch-wide by 20-inch-deep by 90-inch-high indoor-type structure with 600-, 800-, or 1000-ampere horizontal buses and 300-ampere vertical buses. Busbar bracing is provided for 25,000 asymmetrical amperes as standard.

SPACE FACTORS AVAILABLE FOR DISCONNECTS OR STARTERS IN MAIN FEED-IN SECTION			
INCOMING FEED, AMPERES	MAIN LUGS	MAIN BREAKER	MAIN SWITCH
200	5 1/2	—	4
225	—	5	—
400	5 1/2	4 1/2	2 1/2
600	5 1/2	4	1 1/2
800	11	0	0
1000	11	0	0

The labor for installation of this vertical section is based upon its being received from the factory with main lugs, breaker, or switch installed, moving in to location under favorable conditions, and anchoring to the floor. Material costs are based upon the contractors' net book prices.

COST OF 20" DEEP MOTOR CONTROL CENTER VERTICAL SECTION FOR INCOMING MAIN

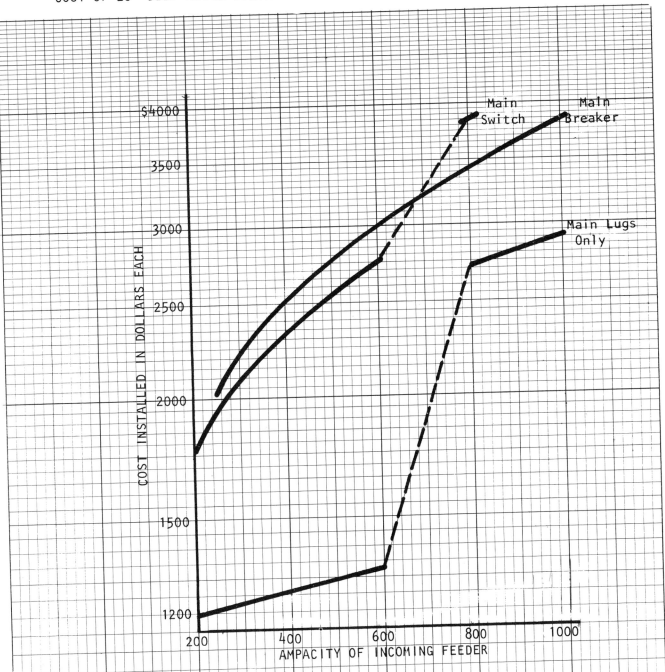

M.C.C. - 20" DEEP VERTICAL SECTION FOR INCOMING FEEDER

G

G-2

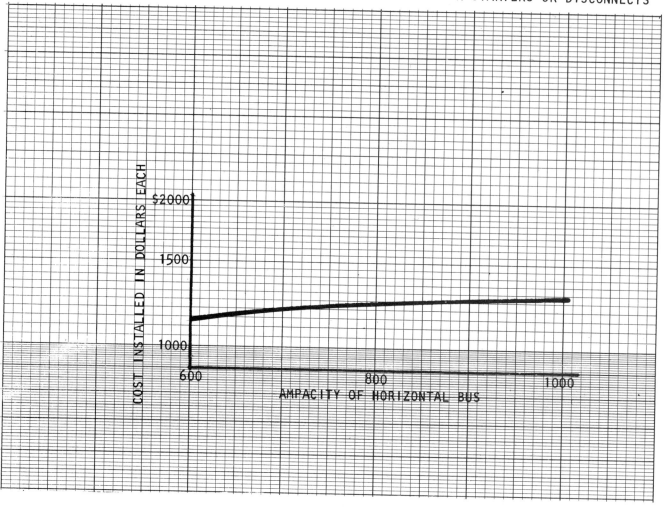

G-25

This vertical section is a 20-inch-wide by 20-inch-deep by 90-inch-high structure with 600-, 800-, or 1000-ampere horizontal bus and a 300-ampere vertical bus. Buses are aluminum and are braced for 25,000 asymmetrical amperes. These sections have 6 1/2 usable space factors for disconnects or starters, allowing wiring accessibility at the bottom. Vertical wiring troughs are also provided.

The labor for installation of the vertical section is based upon moving into location under favorable conditions and anchoring to the floor. Material costs are based upon the contractors' net book prices.

COST OF M.C.C. 600-V BRANCH FEEDER FUSIBLE SWITCHES

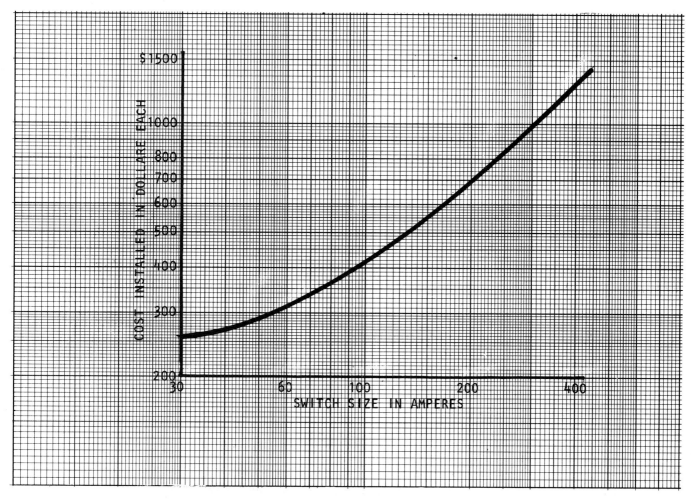

The costs shown for the switch/fuse branch feeder disconnects consist of the published contractors' book prices for the units only. Costs for the vertical section are not included with these costs; they are listed separately for the vertical section only.

These costs also include a conduit termination into the center and 5 feet of three-conductor wire sized to match the disconnect. Switch costs also include 600-volt Fusetrons.

SPACE FACTOR REQUIREMENTS	
SWITCH SIZE	SPACE FACTOR
30/3	1
60/3	1
100/3	1
200/3	2 1/2
400/3	4

The costs shown for the circuit-breaker branch circuit disconnects consist of the published contractors' book prices for the units only. Costs for the vertical section are not included with these costs; they are listed separately for the vertical section only.

These costs include a conduit termination into the center and 5 feet of three-conductor wire sized to match the disconnect.

SPACE FACTOR REQUIREMENTS	
CIRCUIT-BREAKER SIZE	SPACE FACTOR
50/3	1
100/3	1
225/3	1 1/2
400/3	2

COST OF M.C.C. 600-V BRANCH FEEDER CIRCUIT BREAKERS

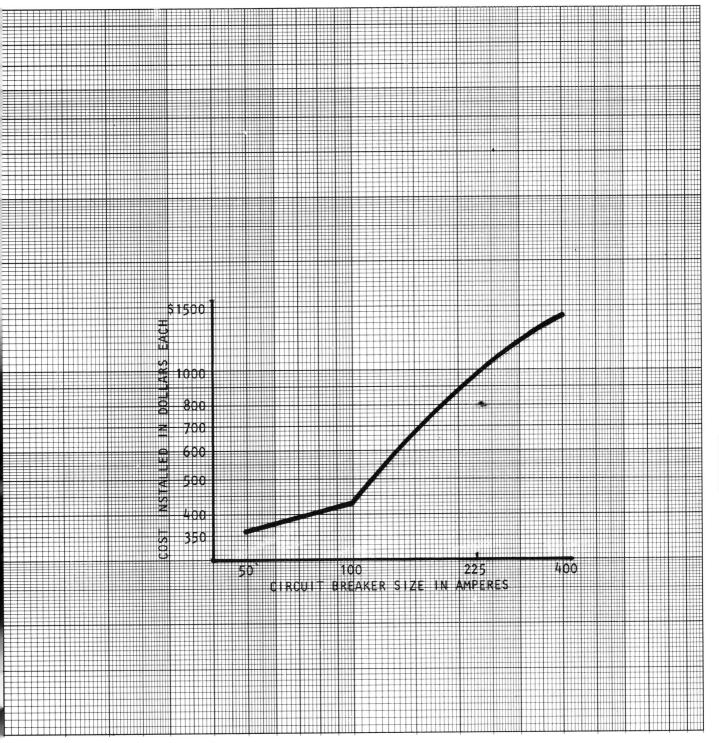

The costs shown for the full-voltage combination starters are based upon the contractors' published book prices. The control center components used in establishing these graphs are based upon Square D Model 4 motor control centers.

Since Class 1, type B, wiring is the most popular, prices are based upon it. With independent wiring for each unit, the wires terminate at terminal blocks near the bottom of each unit. No intercontrol wiring is provided.

A space factor is approximately 12 inches in height.

Included in the costs shown are the following:

1. Combination starter with circuit-breaker type of disconnect.
2. Start-stop switch and pilot light in cover.
3. Two separate interlock switches, one N.O. and one N.C.
4. Control transformer.
5. Overload relays in starter.
6. Conduit terminal into section.
7. Sufficient conductor of size required.

	FULL-VOLTAGE STARTER (FVNR)				
NEMA SIZE	SPACE FACTOR		HORSEPOWER RANGE		
	Circuit Breaker	Fusible Switch	200-Volt Motor	230-Volt Motor	460/575-Volt Motor
1	1	1	0-7 1/2	0-7 1/2	0-10
2	1	1	8-10	8-15	11-25
3	1 1/2	1 1/2	11-20	16-25	26-50
4	2	3 1/2	21-40	26-50	51-100
5	4	5 1/2	41-75	51-100	101-200

COST OF M.C.C. FULL-VOLTAGE NONREVERSING STARTER - 1 SPEED

COST INSTALLED IN DOLLARS EACH

STARTER SIZE BY NEMA SIZE

The costs shown for the full-voltage combination starters are based upon the contractors' published book prices. The control center components used in establishing these graphs are based upon Square D Model 4 motor control centers.

Since Class 1, type B, wiring is the most popular, prices are based upon it. With independent wiring for each unit, the wires terminate at terminal blocks near the bottom of each unit. No intercontrol wiring is provided.

A space factor is approximately 12 inches in height.

Included in the costs shown are the following:
1. Combination starter with circuit-breaker type of disconnect.
2. Start-stop pushbutton and pilot light in cover.
3. Two separate interlock switches, one N.O. and one N.C.
4. Control transformer with fused secondary.
5. Overload relays in starter.
6. Conduit terminal into vertical section.
7. Sufficent conductor of size required.

	FULL-VOLTAGE STARTER (FVR)				
NEMA SIZE	SPACE FACTOR		HORSEPOWER RANGE		
	Circuit Breaker	Fusible Switch	200-Volt Motor	230-Volt Motor	460/575-Volt Motor
1	1 1/2	1 1/2	0-7 1/2	0-7 1/2	0-10
2	2	2	8-10	8-15	11-25
3	2 1/2	2 1/2	11-20	16-25	26-50
4	3	4	21-40	26-50	51-100
5	6 1/2	—	41-50	51-60	101-125

COST OF M.C.C. FULL-VOLTAGE REVERSING STARTER - 1 SPEED

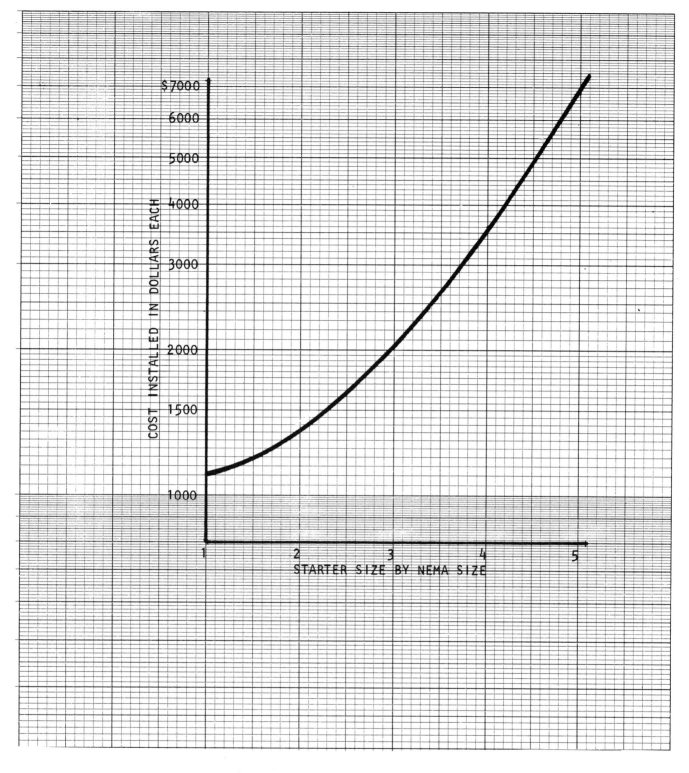

The costs shown for the full-voltage two-speed starters, are based upon the contractors' published book prices. The control center components used in establishing these graphs are based upon *Square D Model 4* motor control centers.

Since Class 1, type B, wiring is the most popular, prices are based upon it. With independent wiring for each unit, the wires terminate at terminal blocks near the bottom of each unit. No intercontrol wiring is provided.

A space factor is approximately 12 inches in height.

Included in the costs shown are the following:

1. Combination starter with circuit-breaker type of disconnect.
2. High-low-stop pushbuttons and two pilot lights in cover.
3. Two separate interlock switches, one N.O. and one N.C.
4. Control transformer with fused secondary.
5. Overload relays in starter.
6. Conduit terminal into vertical section.
7. Sufficent conductor of size required.

	FULL-VOLTAGE STARTER—TWO-SPEED MOTOR				
NEMA SIZE	SPACE FACTOR		HORSEPOWER RANGE		
	Circuit Breaker	Fusible Switch	200-Volt Motor	230-Volt Motor	460/575-Volt Motor
1	2	2	0-5	0-5	0-7 1/2
2	2	2	6-7 1/2	6-10	8-20
3	2 1/2	2 1/2	8-20	11-25	21-40
4	3	4 1/2	21-30	26-40	41-75

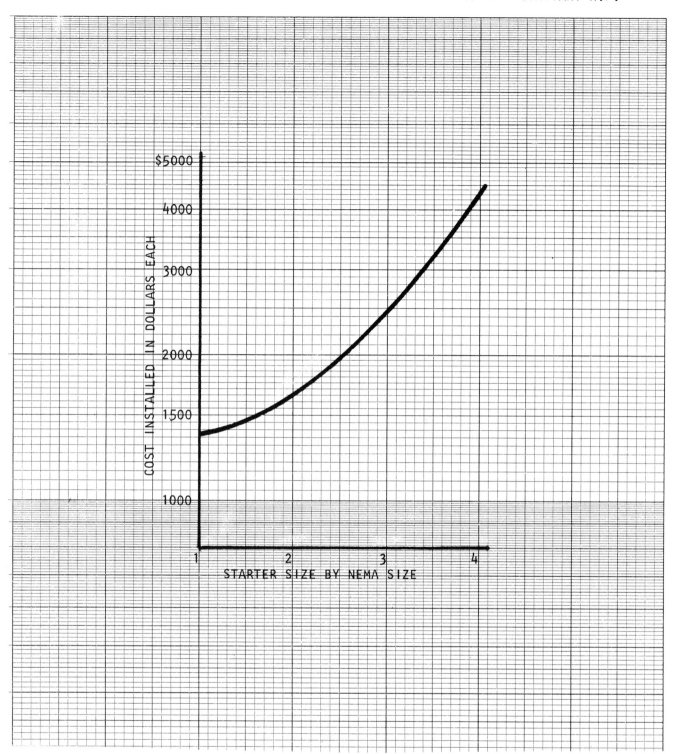

G-30

The costs shown for the reduced-voltage starters are based upon the contractors' published book prices. The control center components used in establishing these graphs are based upon Square D Model 4 motor control centers.

Since Class 1, type B, wiring is the most popular, prices are based upon it. With independent wiring for each unit, the wires terminate at terminal blocks near the bottom of each unit. No intercontrol wiring is provided.

A space factor is approximately 12 inches in height.

Included in the costs shown are the following:

1. Combination starter with circuit-breaker type of disconnect.
2. Start-stop pushbutton and pilot light in cover.
3. Two separate interlock switches, one N.O. and one N.C.
4. Control transformer with fused secondary.
5. Overload relays in starter.
6. Conduit terminal into vertical section.
7. Sufficent conductor of size required in center.

REDUCED-VOLTAGE STARTER—AUTOTRANSFORMER					
NEMA SIZE	SPACE FACTOR		HORSEPOWER RANGE		
	Circuit Breaker	Fusible Switch	200-Volt Motor	230-Volt Motor	460/575-Volt Motor
2	4 1/2	4 1/2	0-10	0-15	0-25
3	5 1/2	5 1/2	11-20	16-25	26-50
4	6	7	21-40	26-50	51-100

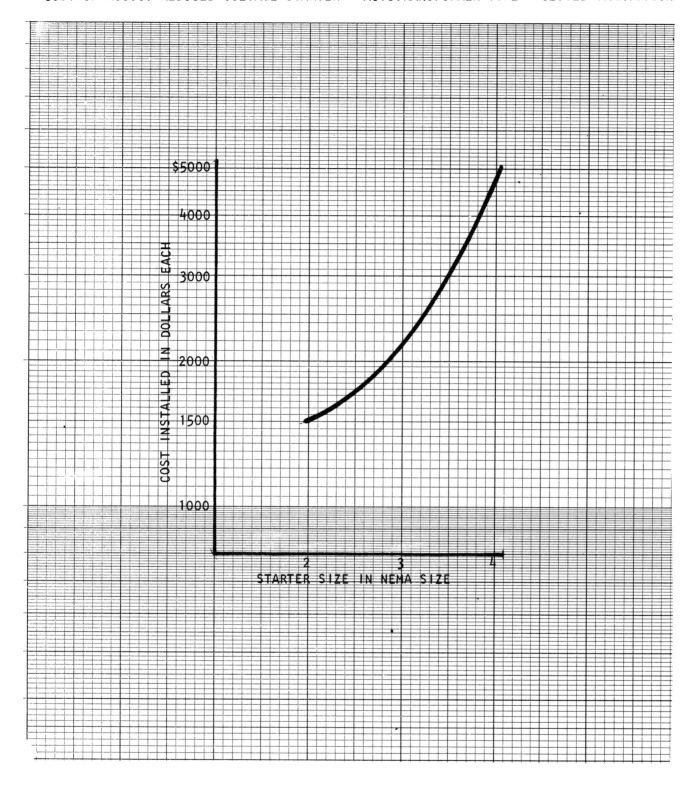

G

G-31

The costs shown for the reduced-voltage starters are based upon the contractors' published book prices. The control center components used in establishing these graphs are based upon Square D Model 4 motor control centers.

Since Class 1, type B, wiring is the most popular, prices are based upon it. With independent wiring for each unit, the wires terminate at terminal blocks near the bottom of each unit. No intercontrol wiring is provided.

A space factor is approximately 12 inches in height.

Included in the costs shown are the following:

1. *Combination starter with circuit-breaker type of disconnect.*
2. *Start-stop pushbutton and pilot light in cover.*
3. *Two separate interlock switches, one N.O. and one N.C.*
4. *Control transformer with fused secondary.*
5. *Overload relays in starter.*
6. *Conduit terminal into vertical section.*
7. *Sufficient conductor of size required in center.*

REDUCED-VOLTAGE STARTER – PART-WINDING (TWO-STEP)					
NEMA SIZE	SPACE FACTOR		HORSEPOWER RANGE		
	Circuit Breaker	Fusible Switch	200-Volt Motor	230-Volt Motor	460/575-Volt Motor
1-PW	2 1/2	2	0-10	0-10	0-15
2-PW	2 1/2	2 1/2	11-20	11-25	16-40
3-PW	3	4	21-40	26-50	41-75
4-PW	4 1/2	—	40-75	51-75	76-150
5-PW	7	—	76-125	76-150	151-300

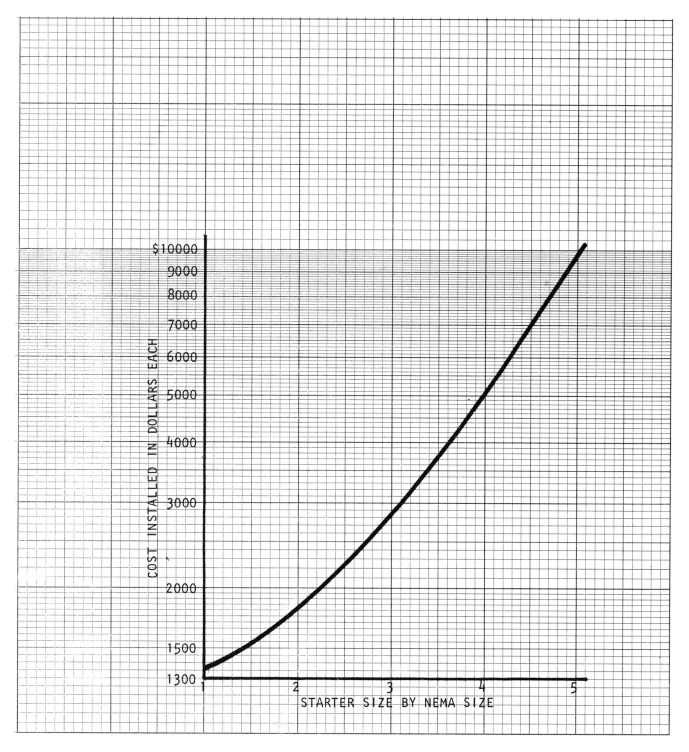

G

G-32

Transformers

DRY TYPE

Single Phase—General Purpose —0.05-3.0 kVA   H-1

Single Phase—General Purpose—3.0-167 kVA   H-2

Single Phase—Buck-Boost—0.25-3.0 kVA   H-3

Single Phase—Isolating Type—1.5-25 kVA   H-4

Three Phase—General Purpose—3.0-500 kVA   H-5

Three Phase—Autotransformer—30-300 kVA   H-6

Three Phase—Isolating Type—15-75 kVA   H-7

OIL TYPE

Single Phase—Pole Type—10-100 kVA   H-8

Three Phase—Pad-Mounted Type—75-1000 kVA   H-9

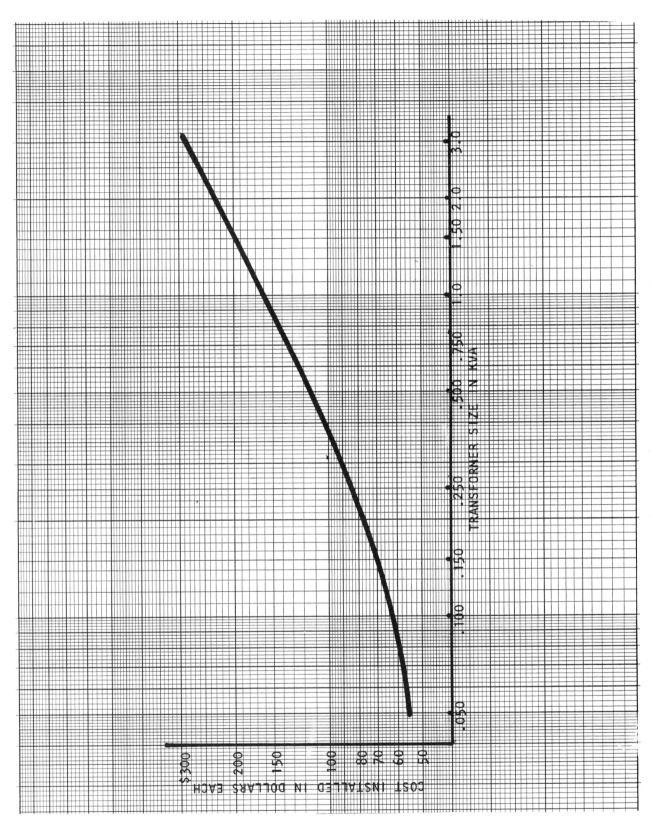

The costs shown for these single-phase dry transformers consist of the published contractors' book prices and are shown in the Square D Co. catalog. These costs include anchoring to a masonry wall with toggle bolts and installing the required conduit and wire terminals.. The transformers are the Sorgel Series for general light and power trand are rated 240/480, 120/240, with no taps and are for indoor or outdoor use.

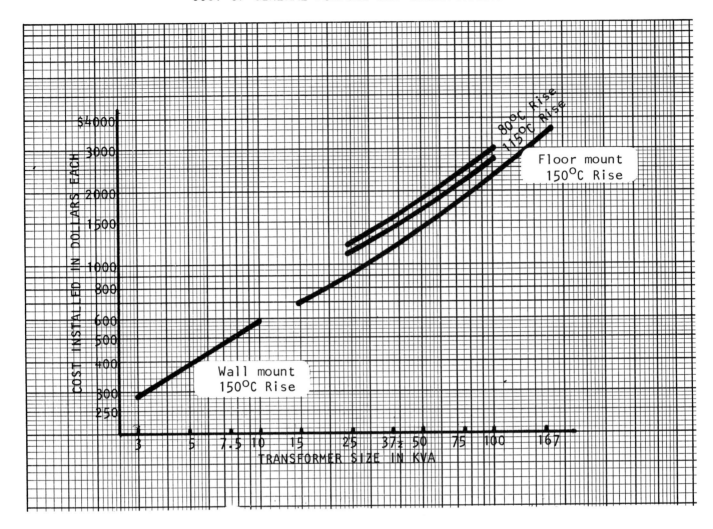

The costs shown for the single-phase dry transformers consist of the published contractors' book prices as listed in the Square D Co. catalog. These costs include anchoring to a wall or floor as required and installing the required conduit and wire terminals. The transformers priced are as described below:

SINGLE-PHASE	LIGHT AND POWER	480-120/240
.05 - 15	Wall-Mounted	
25 - 167	Floor-Mounted 6 - 2 1/2%, 2AN, and 4BN	

Note: 2AN means two taps for a supply voltage above normal, and 4BN means four taps for supply voltages below normal.

H-2

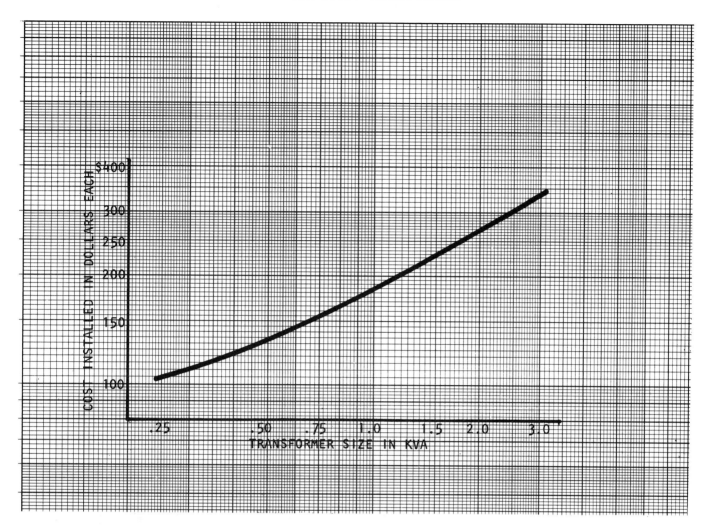

The costs shown for these single-phase dry buck-boost transformers
consist of the published contractors' book prices as shown in the Square D
Co. catalog. These costs include anchoring to a masonry wall with toggle
bolts and installing the required conduit and wire terminals.
Transformers ratings are:

> 120/240–12/24
> 120/240–16/32

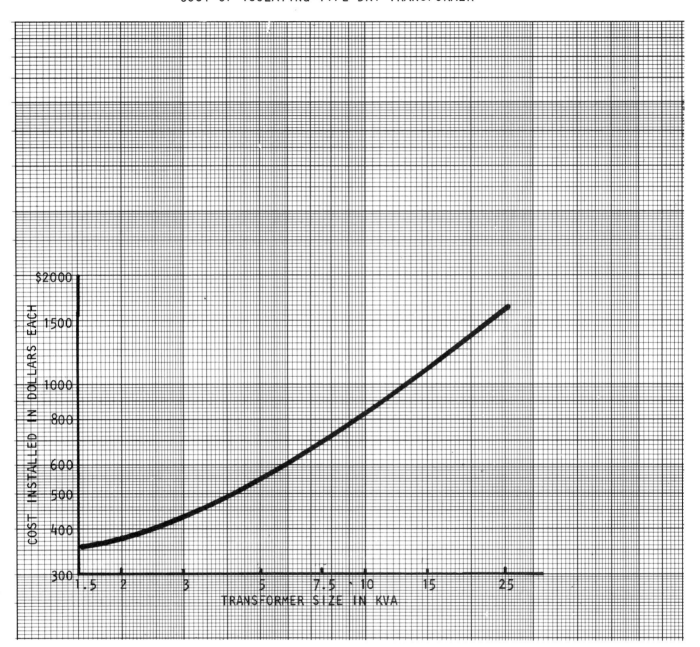

The costs shown for these isolating transformers consist of the published contractors' book prices as shown in the Square D Co. catalog. These costs include anchoring to a wall or floor as required and installing the required conduit and wire terminals.

While any two-winding transformer is an isolating transformer, these particular transformers are designed to reduce the "electrical hash" that comes in through the utility system. They are often used with electronic motor controls, x-ray machines, and computers.

ISOLATING-TYPE DRY TRANSFORMER - SINGLE-PHASE

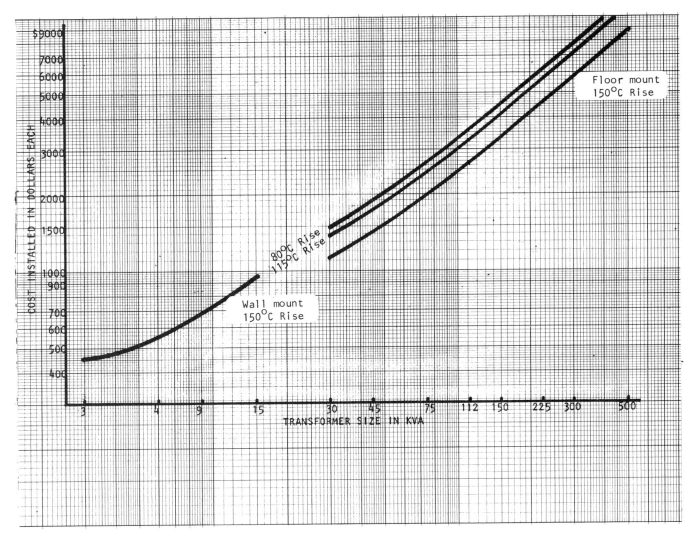

The costs shown for the three-phase dry transformers consist of the published contractors' book prices as shown in the Square D Co. catalog. The costs include anchoring to a wall or floor as required and installing required conduit and wire terminals. The transformers priced are as described below:

THREE-PHASE   LIGHT AND POWER   480-208Y/120

  3 - 15   Wall-mounted 2 - 5% BN
  30 - 500   Floor-mounted 6 - 2 1/2%, 2AN, and 4BN

Note: 2AN means two taps for a supply voltage above normal, and 4BN means four taps for supply voltages below normal.

GENERAL-PURPOSE DRY TRANSFORMER - THREE-PHASE

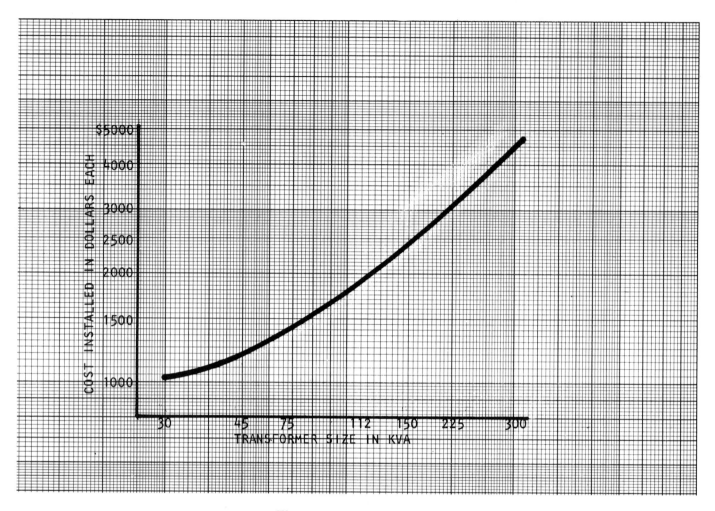

The costs shown for these three-phase autotransformers consist of the published contractors' book prices as shown in the Square D Co. catalog.

Where they are not restricted by local ordinances or where isolation is not required by certain equipment, they are economical and quieter and have better voltage regulation.

These transformers are rated at 480Y/277 volts, three-phase, four-wire, to 208Y/120 volts and cannot be used on a 480-volt, three-wire system.

THREE-PHASE AUTOTRANSFORMER - DRY TYPE

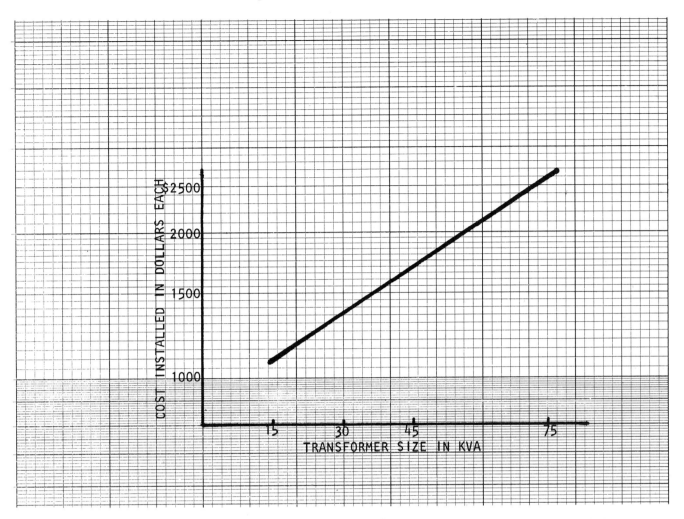

The costs shown for these isolating transformers consist of the published contractors' book prices as shown in the Square D Co. catalog. These costs include anchoring to a wall or floor as required and installing the required conduit and wire terminals.

While any two-winding transformer is an isolating transformer, these particular transformers are designed to reduce the "electrical hash" that comes in through the utility system. These transformers are frequently used to supply electronic equipment.

H-7

H-8

The costs shown for the pole-mounted transformers consist of the contractors' published book prices and are as described below.

The pole-type transformers represented are two bushing transformers. The voltages used were 7200/12,470Y to 120/240 with four 2 1/2% taps below normal. Reference is General Electric type HSBA.

Installation costs include mounting the transformer and making high- and low-voltage connections.

SINGLE-PHASE TRANSFORMERS - POLE TYPE - OIL-FILLED

COST OF PAD-MOUNTED TRANSFORMERS

COST INSTALLED IN DOLLARS EACH

$15000

10000

8000

6000

5000

4000

75   100   150   200   300   500   1000

SIZE OF TRANSFORMER IN KVA

The costs shown for the pad-mounted transformers consist of the
contractors' published book prices and the items described below.

The pad-mounted transformers shown are considered to be single three-
phase transformers of the sizes shown, with high-voltage terminations
and secondary wire terminals sized to match the transformer ampacity.
Also included is a 6-inch concrete base with rebar as a transformer
foundation. The transformers referenced are GE Compads found in
Catalog Section 5434, page 2, are rated 12,470 Grd/7200 to 208
GrdY/120, and have four 2 1/2% taps below normal.

Auxiliary Systems

TELEPHONE SYSTEM

Telephone Cabinets (For Outlets, See Branch Circuits)   I-1
Undercarpet Communications System Components   I-2
Undercarpet Communications Cable   I-3, I-4
Telephone Plenum Cable   I-5

SIGNALING SYSTEM

Low-Voltage Transformers and Pushbuttons   I-6
Low-Voltage Audible Signals, Bells, and Buzzers   I-7
High-Voltage Pushbuttons, Bells, and Buzzers   I-8
Industrial/Institutional Audible Signals   I-9

CLOCK/PROGRAM SYSTEM

Indicating Clocks, Synchronous-Wired, and Electronic   I-10
Master Time Control and Time-Tone Boxes   I-11
Frequency Generators, Relays, and Accessories   I-12

FIRE ALARM SYSTEM

Control Panels   I-13, I-14
Battery-Power Pack, Annunicator Panel, Remote-Receiving Panel   I-15
Manual Stations, Heat Detectors, Sprinkler Flow, and
    OS&Y (Tamper) Switch   I-16
Smoke Detectors   I-17
Door Holder, Horn/Speaker, Horn/Light, and Phone   I-18
Fire Alarm Plenum Cable   I-19

NURSE CALL SYSTEM

Automatic Full-Feature Control Unit, Staff, Duty,
    Staff Locator Stations   I-20
Automatic Full-Feature Central Equipment Cabinet, Patient Station,
    and Emergency Call Station   I-21
Economy System Control Unit, Patient Station, Staff/Duty Station,
    Power Supply/Central Equipment & Cabinet   I-22
Dome Light and Cables   I-23

I

SOUND SYSTEM

Master Unit, Call-in/Privacy Switch   I-24
Speakers, Horns, and Volume Control   I-25
Microphone Receptacles, Mixers, and Amplifiers   I-26
Microphones   I-27
Antenna, Tuner, and Cables (Pulled-in Conduit)   I-28
Sound System Plenum Cable   I-29

EMERGENCY CALL SYSTEM

Room Station, Door Release, Corridor Light, Bell, and Transformer   I-30
Smoke Detector, Remote-Indicator Emergency Phone, Power Supply,
   Master Annunciator   I-31

APARTMENT INTERCOM SYSTEM

Apartment Speaker, Door Release, Power Supply, Amplifier   I-32
Directory, Master Unit, and Control Unit   I-33

MASTER TELEVISION ANTENNA SYSTEM

Antennas, Amplifiers, and Signal Splitters   I-34
Outlets, Baluns, Cable, and Cable Terminators   I-35

4" Deep Flush Cabinets	
Size	Cost Installed
12" x 24"	$159.
18" x 30"	199.
24" x 30"	278.
30" x 30"	328.
30" x 36"	362.
36" x 48"	662.
3/4" Plywood Board	
4' x 8'	$54.
4' x 4'	38.

The costs shown for the telephone cabinets consist of the published contractors' book prices for cabinets of the sizes shown, which are related to frontal area and a 4-inch-deep cabinet. The plywood board is finished with black insulating varnish. It is further assumed that the cabinet is mounted flush in a masonry wall and has 1 1/4- to 2 1/2-inch GRC conduit terminal included. Terminal size depends upon cabinet size.

I-1

Description		Installed Cost
Wall Transition Box - for 25 pair round cable to 25 pair flat cable	Without rack	$32.25
Wall Transition Box - for 3 or 4 pair round cables to two 3 or 4 pair flat cables.	With rack and 1 - 3 or 4 pair double connector assy. For two telephones.	63.75
	With rack and 2 - 3 or 4 pair double connector assy. For four telephones.	87.75
	With rack and 3 - 3 or 4 pair double connector assy. For six telephones.	110.00
	With rack and 4 - 3 or 4 pair double connector assy. For eight telephones.	132.25
Wall Transition Box - for 25 pair round telephone cable to 3 or 4 pair flat cables.	With rack and adapter for 25 pair cable and connectors for 3 pair flat cables.	71.50
Telephone Pedestal	25 Pair flat to 25 pair telephone cable	42.75
	For 1 - 3 or 4 pair telephone connection	59.25
	For 2 - 3 or 4 pair telephone connections.	66.75
Call Director Pedestal		51.00

The costs shown for these undercarpet communications system components consist of published contractors' prices and the labor for installation.

I-2

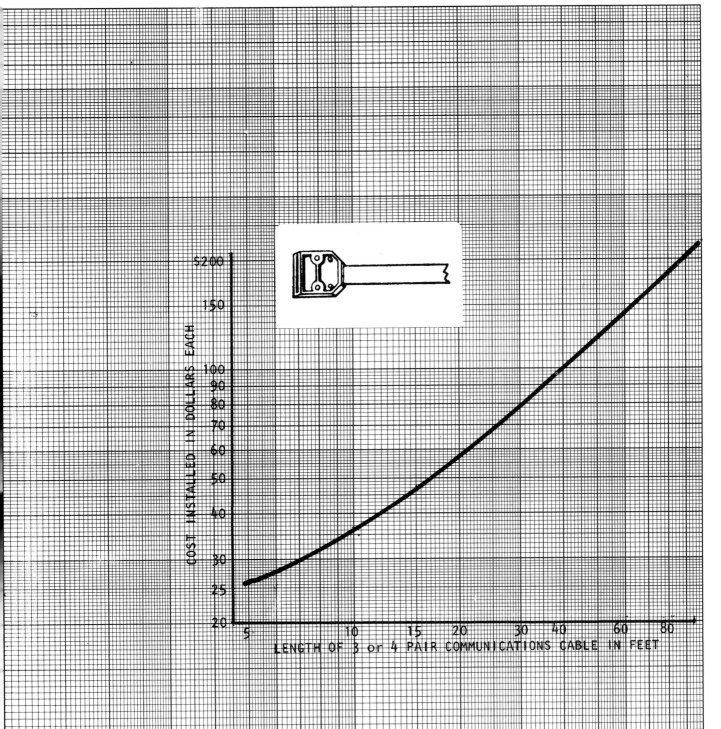

COST INSTALLED IN DOLLARS EACH

LENGTH OF 3 or 4 PAIR COMMUNICATIONS CABLE IN FEET

The costs shown for undercarpet communications cable consist of the published contractors' book price for material and the labor for installation.

I-3

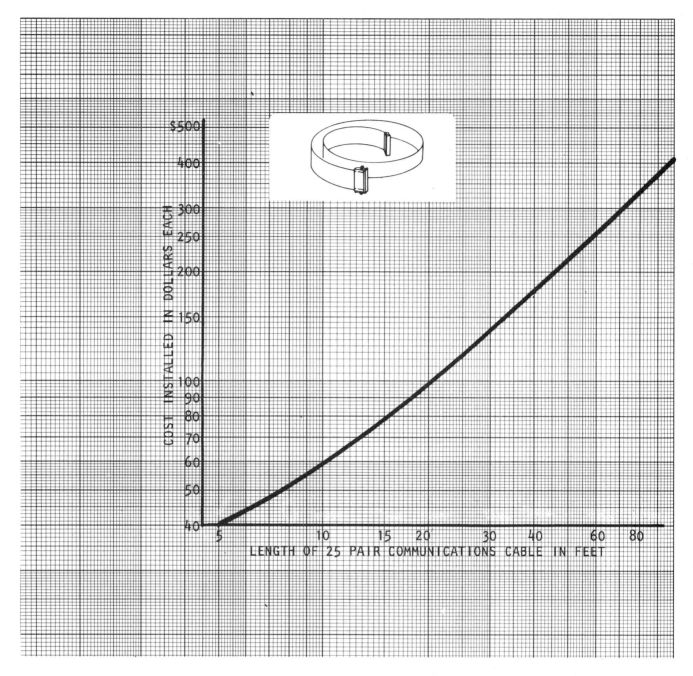

I-4

The costs shown for the undercarpet communications cable consist of the published contractors' price for material and the labor for installation.

25 PAIR UNDERCARPET COMMUNICATIONS CABLE - MADE-UP ENDS

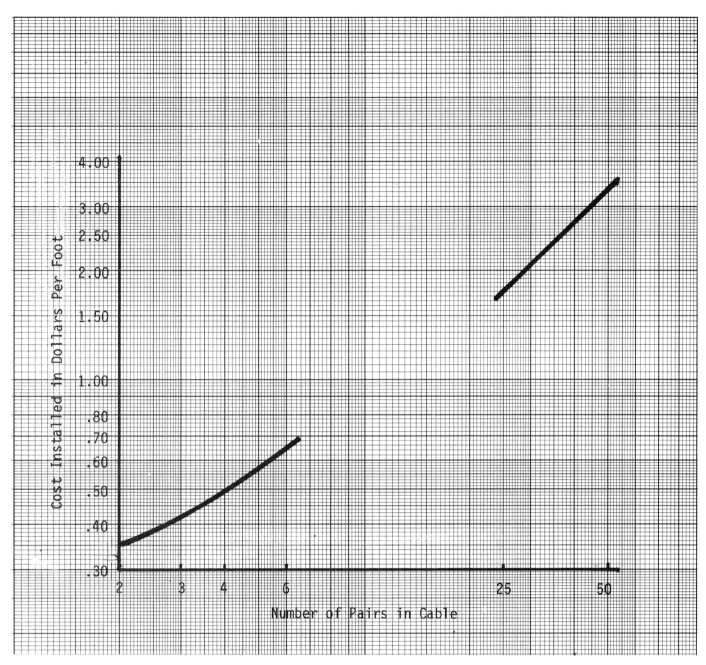

The costs shown for high-temperature telephone cables consist of the published contractors' price for Teflon jacketed cable approved for use in environmental air plenums. Included is the labor for installing cable without conduit in accessible ceiling cavities.

I-5

LOW VOLTAGE (below 50 volts)

Type of Unit	Manufacturer	Amount
Transformer	Edwards-No.592 10 VA - 8, 16, 24V	$36.00
Transformer Same as above	Edwards-No.598 30VA - 8, 16, 24V	44.00
Transformer	Edwards 50VA - No.88-50, 12, 24V 100VA - No.88-100, 12, 24V 250VA - No.88-250, 12,16,20,24V	73.00 92.50 175.00
Pushbutton	Edwards Interior to 48volts No.695-11 w/ s.s. plate & No.147 Exterior to 48volts No.1786-C	55.00 44.00

I-6

The costs shown for this light commercial signaling equipment consist of the published contractors' book prices. The interior-type pushbutton 695-11 also includes a stainless-steel plate upon which to mount the pushbutton. The 1786-C pushbutton has a threaded conduit entrance on the back plate.

COST OF SIGNALING EQUIPMENT
Light Commercial

LOW VOLTAGE(below 50 volts)

Type of Unit	Manufacturer		Amount
Flushcall Series	Edwards	Catalog Number	
	Powacall transformer 590&593		$58.00
	Ringacall Bell	660	56.00
	Buzzacall Buzzer	661	56.00
	Tucall (bell ¢ buzzer) 662		62.00
	Melocall	663	58.00
	Togelpush (pushbutton) 664		52.00
Bell, Buzzer. or Combination	Edwards		
	6 VAC Bell	720	14.00
	6 VAC Buzzer	725	13.45
	6 VAC Bell/Buzzer	730	18.50
Bell	Edwards		
	8 VAC Bell	55-4"	27.00
Buzzer	Edwards		
	8 VAC Buzzer	15-0	21.50

I-7

The costs shown for these audible signals consist of the published contractors' book prices and are based upon the Edwards products. The costs include the labor and materials for either flush or surface type of mounting as required.

COST OF SIGNALING EQUIPMENT
Light Commercial

HIGH VOLTAGE (50 - 125 volts)

Type of Unit	Manufacturer	Amount
Pushbuttons	Edwards Interior No.850 with no. 147 s.s. plate	$63.50
Pushbuttons Same as above	Edwards Exterior No. 852 with stainless steel plate	69.00
Pushbuttons	Edwards Outdoor No.1785, Weatherproof, for 1/2" conduit mounting	80.00
Bells, Buzzers	Edwards Bells No. 340A-4N5 120V Buzzers No. 340A-N5 120V	96.00 87.00

The costs shown for the light commercial high-voltage type of pushbutton consist of the published contractors' book prices. The outdoor pushbutton (#1785) is provided with a 1/2-inch conduit entrance.

HIGH-VOLTAGE PUSHBUTTONS, BELLS, & BUZZERS

COST OF AUDIBLE SIGNALS-INDUSTRIAL/INSTITUTIONAL
For fire alarm and programs, etc.

Description	Cost Installed in Dollars Each		
Bells, Horns, Sirens, and Chimes	4" Diameter	6" Diameter	10" Diameter
Single Stroke Bells			
Surface Mounted	$ 88.	$105.	$125.
Flush Mounted	134.	150.	---
Weatherproof	121.	136.	151.
Vibrating Bells			
Surface Mounted	87.	104.	125.
Flush Mounted	134.	150.	---
Weatherproof	121.	138.	150.
Cast Aluminum Guard	N/A	46.	55.

Description	Alternating Current	Direct Current
Federal No. 350 & 450 Grill Model Horns		
Surface Mounted	$ 86.	$113.
Flush Mounted	122.	150.
Weatherproof	115.	143.
Semi-flush	125.	152.
Federal No.350 & 450 +PR Single Projector Horns		
Surface Mounted	106.	134.
Flush Mounted	143.	170.
Weatherproof	134.	161.
Federal No.350 $ 450+2PR Double Projector Horns		
Surface Mounted	111.	151.
Flush Mounted	148.	165.
Weatherproof	140.	157.

Description	Cost
Federal Model A Siren	$408.
Edwards No. 338 Single Stroke Chime	110.

The costs shown for these audible signals consist of the published contractors' book prices and are based upon Federal Sign and Signal Corporation products, which are used extensively in OEM equipment. The costs include labor and material for either flush or surface type of installation.

The costs shown for these clocks consist of the published contractors' book prices for the units described and are as manufactured by Simplex Time Recorder Company. Note that the 15-inch clocks are not available for mounting in time-tone units.

Included: (material and labor)

1. Clock with recessed clock receptacle and box in masonry wall for surface-mounted unit.
2. Clock with special recessed back box and conduit terminals for semiflush mounting.
3. Installing and connecting in time-tone units or special back box as required.

Excluded:

1. Circuit wiring to nearest power outlet.
2. Buzzers in clock are not included but will fit in the back box if required.

COST OF CLOCK/PROGRAM SYSTEM CLOCKS

Clocks		Cost Installed in Dollars Each					
		Type of System					
Mounting	Clock Face	Indicating Only No Correction		Synchronous Wired With Individual Automatic Correction		Electronic With Individual Automatic Correction	
		Type of Mounting		Type of Mounting		Type of Mounting	
		Individual	In Time Tone Units	Individual	In Time Tone Units	Individual	In Time Tone Units
Surface Mounting	Round 9"	$139.	N.A.	$142.	N.A.	$163.	N.A.
	12"	108.		140		166	
	15"	135.		155.		180.	
	Square 12"	136.		155.		184.	
Semi-Flush Mounting	Round 9"	145.	$109.	150.	$114.	163.	$127.
	12"	113.	81.	153.	118.	166.	130.
	15"	137.	N.A.	165.	N.A.	194.	N.A.
	Square 12"	137.	126.	162.	154.	190.	154.
Wall or Ceiling Mounted Dbl Dial	Round 12"	280.	N.A.	356.	N.A.	374.	N.A.
	Square 12"	374.	N.A.	430.	N.A.	430.	N.A.
Flush	Exec. 12"	461.	N.A.	508.	N.A.	508.	N.A.
	Series 15"	559.		605.		601	
	18"	638.		685.		685.	
Celestra Digital	Surface	538.	N.A.	541.	N.A.	611.	N.A.
	Semi-Flush	564.		567.		638.	
	Dbl Dial	1170.		1170.		1317.	

*The costs for the time-tone units consist of the published contractors'
book prices for the units described and manufactured by Simplex Time
Recorder Company.*

Included: (material and labor)

1. *Back box for semiflush mounting in masonry wall.*
2. *EMT terminals and speaker grille as shown.*

Excluded:

1. *Clocks and speakers.*
2. *Wire and connecting thereto.*

*The costs shown for the master time center units consist of the published
contractors' book prices for the units described and are as manufactured
by Simplex. Each unit has the capability of running 24 hours during
power outage before losing time. The units are the modern design for
semiflush mounting.*

Included: (material and labor)

1. *Clock.*
2. *Back box with EMT terminals.*

COST OF CLOCK/PROGRAM SYSTEM

TIME-TONE UNITS (Clocks & Speakers not included)

Type of Unit	All by Simplex	Cost Installed In Dollars Each
	T Series: For 12" round or square clock and rectangular speaker grille	$109.
	TC Series: For round clocks with round speaker grille	
	For 9" clock	109.
	For 12" Clock	112.
	TD Series: For square clock with square speaker grille	
	For 12" Clock	122.
Master Time & Program Center	Type 2351: For time control only - synchronous wired or electronic	1240.
	Type 2350: For time and program control - synchronous wired or electronic	2913.
	Type 2400: For time, program, and building management - synchronous wired or electr.	3992.

	Description	Cost Installed Each
FREQUENCY GENERATOR FOR ELECTRONIC SYSTEM	Simplex - type 2850 Static Type, 4 Output Frequencies 300 or 600 volt System	$1953.
CODED RELAYS	Simplex - type 2801 Coded Relay for Control of Audible Signals and Utilities Wired System Electronic System	 536. 567.

CLOCK ACCESSORIES	Buzzers
	
	  Mounted in Clock Back Box
	  Individual Flush Mounting
	
	Clock Guards (thru 15 inch)
		
		26.
		74.
		
		113.

The costs shown for these static and rotating types of frequency generators and clock and program system relays consist of the published contractors' book prices and the labor for moving to location and mounting to wall or floor as required.

Included: (material and labor)
* Connection wiring in control cabinet*
* Mounting and fastening in place*

Excluded:
* Disconnecting switches or breakers*
* Conduit and wire to or from control cabinet*

I-12

Type of Unit	Description	Amount
	Simplex 2001 Series	
	Flush or Surface Mounted - Panel includes 2 zones, city relay, one **supervised horn circuit (6 horns/cct)** battery charger and 9 AH battery	$1705.
	Add For:	
	Each Additional Zone	149.
	Each Additional Horn Circuit	122.
	Fan Shutdown Relay	75.
	Aux. Function Module (for door hldrs.)	122.
	Additional Power Supply	222.

The Simplex 2001 Series Fire Alarm control panel is the new solid-state modular type—24-volt, DC-operated. This basic unit contains two zones, one city relay and one supervised horn circuit, battery, and charger.

Features included:

a. *Power to all zone-alarm-initiating circuits.*
b. *May be used for coded or noncoded operation.*
c. *Provides for either one- or two-zone circuits.*
d. *Green power ON lamps.*
e. *One red alarm and one amber trouble lamp per zone.*
f. *Will operate auxiliarized muncipal fire alarm boxes.*
g. *May be connected to operate a shunt-type trip muncipal box or local-energy-type muncipal box.*
h. *Has door-mounted key drill switch which sounds only the alarm signals and prevents activation of the muncipal box.*
i. *Transfers to a 24-volt battery in case of AC power failure.*
j. *One signal circuit having output of 1.5 amp each.*
k. *Will control 11 bells per circuit rated at 0.13 amp each.*
l. *Will control 5 horns per circuit rated at 0.30 amp each.*

I-13

Type of Unit	Description	Amount
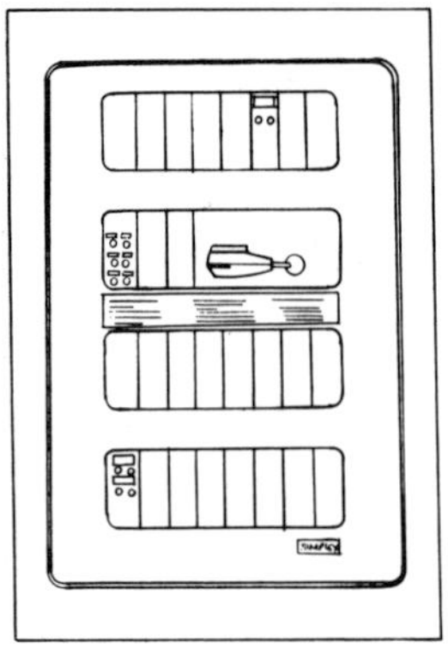	Simplex - 2001-8004 Flush or Surface Mounted Will accommodate a 10 story building zoned by floor	$9325.

This Simplex series fire alarm panel is 24-volt, DC-operated. In addition, this panel provides audio evacuation tone through speakers, public address through speakers, and two-way telephone communication between control panel and remote phone jack locations.

Features included:

    a.   Power to all zone-alarm-initiating circuits.
    b.   May be used for coded or noncoded operation.
    c.   Provides for either one- or two-zone circuits.
    d.   Green power ON lamps.
    e.   One red alarm and one amber trouble lamp per zone.
    f.   Will operate auxilturized muncipal fire alarm boxes.
    g.   May be connected to operate a shunt-type trip muncipal box or a local-energy-type muncipal box.
    h.   Has door mounted key drill switch which sounds only the alarm signals and prevents acitivation of the muncipal box.
    i.   Transfers to a 24-volt battery in case of ac power failure.

I-14

Type of Unit	Description	Amount
Battery Power Pack 	Simplex Cabinet and Battery Charger Surface Mounted with Lead Calcium Batteries 5.2 AH Capacity 8 AH Capacity 18 AH Capacity 33 AH Capacity	$451. 499. 606. 753.
Annunciator Panel	Simplex - 4306-8 Flush type, backlighted, with 2 lamps per zone. 8 Zones	576.
Remote Station Receiving Panel	Simplex - 4293 For one zone (customer) For each additional zone, add	948. 117.

Battery-Power Pack

Monitors voltage condition of wet-cell batteries to supply proper charge. Has manual switched supply and equalized charge to lead-acid batteries when required.

Fire Alarm Annunciator Panel

This annunicator is of the back-lighted type with two lamps per zone for additional reliabilty. This is an eight-zone unit; others are available. Trouble silence, drill, and system reset switches can be included.

Remote Station Receiving Panel

This panel is usually installed at a fire station or central headquarters. It will monitor one or more separate local premises. It contains an alarm and trouble lamp for each premise. It operates a 12-volt DC alarm bell. It requires a separate standby battery. Normal operation is from local power.

I-15

FIRE ALARM BATTERY-POWER PACK, ANNUNCIATOR, & REMOTE-RECEIVING PANELS

Type of Unit	Description	Amount	

Manual Station	Simplex - 2099-9201
	  Surface Mounted
	    1 N.O. Contact
	    2 N.O. Contacts
	
	  Semi-Flush
	    1 N.O. Contact
	    2 N.O. Contacts
		$79.
		91.
		
		64.
		76.	

Manual Station Guard	Simplex	38.	
Automatic Heat Detector	Simplex	With Outlet Box	No Outlet Box
	Fixed Temperature Only Fixed Temperature plus rate of rise Xpl Proof type with either detector	63. 67. 139.	35. 39.
Water Flow Switch	Simplex WF-5	256.	
OS&Y Switch	Simplex Monitor valve position (tamper switch)	172.	

The costs for these fire alarm devices consist of the published contractors' book prices as established by Simplex Time Recorder Company. The costs shown include flush wall or ceiling boxes as indicated, depending upon the type of mounting of the unit, and include the necessary labor for installation as shown.

I-16

FIRE ALARM INITIATING DEVICES

Type of Unit	Description	Amount	
		With Outlet	No Outlet
Smoke Detector Photo Electric Type Ceiling Mounted	Simplex #2098-9603	$133.	$105.
Smoke Detector Ionization Type Ceiling Mounted	Simplex #2098-9508	126.	98.
Smoke Detector - Photo Electric Type - Duct Mounted with Sampling Tubes	Simplex #2098-9633	$227	

The costs shown for these fire detectors consist of the published contractors' book prices as established by Simplex Time Recorder Company. The costs include flush ceiling boxes and labor for installation.

I-17

Type of Unit	Manufacturer	Amount
Magnetic Door Holder 	Simplex Wall Mtd. for Single Door-FM998 Floor Mtd. - Single Door-FM980 Floor Mtd. - Double Door-FM981	$131. 214. 312.
Fire Alarm Speaker 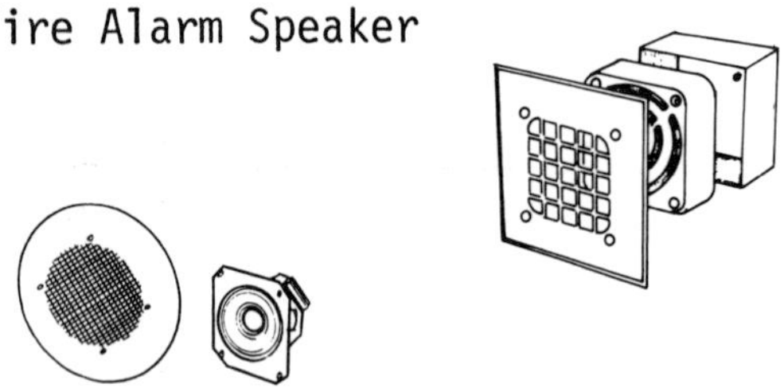	Simplex 2902-9707 Simplex 2902-9704	125. 89.
Fire Alarm Horn with Flashing Light 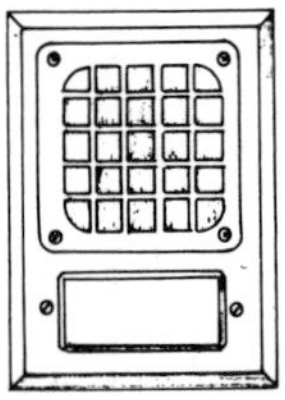	Simplex 2903 series	167.
Emergency Telephone Remote Jack 	Simplex 4590-A	71.
Emergency Telephone 	Simplex 4590-E	61.

The costs shown for these components include back boxes or standard outlet boxes as required and necessary labor for installation.

I-18

FIRE ALARM SYSTEM COMPONENTS

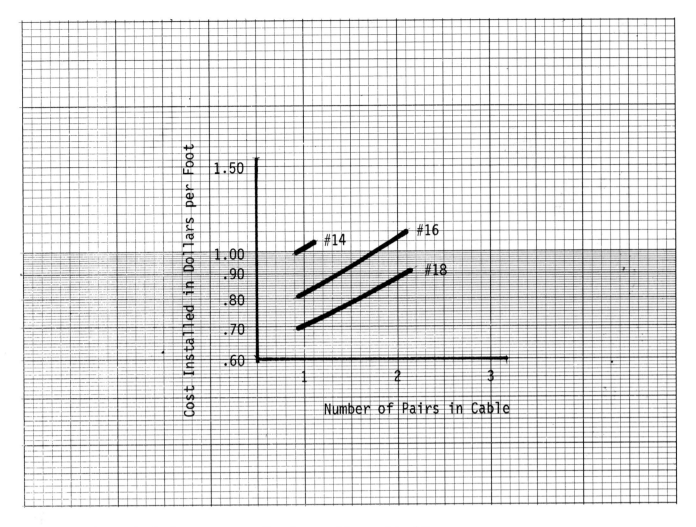

The costs shown for high-temperature fire alarm cable consist of the
published contractors' book price for Teflon-jacketed cable approved
for use in environmental air plenums. Included is the labor for installation
without conduit in accessible ceiling cavities.

I

COST OF AUTOMATIC FULL-FEATURE NURSE CALL SYSTEM COMPONENTS

Type of Unit	Manufacturer	Amount
Nurse Control Station	Rauland Responder III 64 Station	$1402.
Staff Station	Rauland SS100	213.
Duty Station	Rauland DS200	224.
Staff Locator Station	Rauland SRS100	111.

I-20

The costs shown for these nurse call system components consist of the published contractors' book prices and are based upon Rauland products. Included are the necessary installation labor for the equipment and a certain amount for supervision, final checkout, and training of operating personnel.

Type of Unit	Manufacturer	Amount
Central Equipment Cabinet	Rauland Responder III 64 Station	$2810.
Patient Bedside Station	Rauland-Without pillow speakers but with call cords Single Station Double Station	 221. 254.
Emergency Call Station	Rauland #PCS11 pull cord type #PBS11 push button type	 104. 107.

I-21

The costs shown for these nurse call system components consist of the
published contractors' book prices and are based upon Rauland products.
Included are the necessary installation labor for the equipment and a
certain amount for supervision, final checkout, and training of
operating personnel.

Type of Unit	Manufacturer	Amount
Nurse Control Station	Rauland Responder 3000 0-24 Station Desk Type	$773.
Patient Bedside Station	Rauland BS3000	125.
Duty/Staff Station	Rauland SS3000	128.
Power Supply and Central Equipment Cabinet	Rauland NCS3000 for 24 stations	1900.

I-22

The costs shown for these nurse call system components consist of the published contractors' book prices and are based upon Rauland products. Included are the necessary installation labor for the equipment and a certain amount for supervision, final checkout, and training of operating personnel.

COST OF NURSE CALL SYSTEM COMPONENTS

Type of Unit	Manufacturer	Amount
Corridor Dome Light	Rauland	
	#CL7584 Four color signal	$61.
	#CL7582 Two Color Signal	56.
Cable	Rauland	
	2/C	.54
	4/C	.63
	8/C	.78
	12/C	1.04
	15/C	1.39

The costs shown for these nurse call system components consist of published contractors' book prices. Included in the dome light is the labor for installation on an outlet box. The box is included. The cables include the labor for installation in conduit. Refer to the Branch Circuit section for conduit.

I-23

Type of Unit	Manufacturer	Amount

Compact Master Unit	Rauland-Chime/Light Call-In Basic Unit #DIR350LC
	  with 25 selector switches
	  5 watt intercom ampl. and
	  35 watt program ampl. in
	  19" wide cabinet
		$1813.

	Add-On Options	
	14" high aux. cabinet	169.
	Additional 25 selector switch bank	689.
	AM/FM tuner and cassette play.	605.
	Time tone generator	22.
Call-In Switch	Rauland	
	#2304 flush pushbutton	50.
	#2307 flush rocker w/ privacy	55.

The equipment is commonly used in small schools or other buildings where background music distribution, selective paging, and two-way intercommunication are required. This equipment is compatible with speakers and other accessories on following pages. The costs shown include published contractors' book prices including outlet boxes where required. Included are the necessary installation labor for the equipment and a certain amount for supervision, final checkout, and training of operating personnel.

I-24

Type of Unit	Manufacturer	Power	Amount
Trumpet Reproducer	Atlas	Watts	
	Model AP-15C	15	$122.
Wall Speaker	Frasier		
	Model F-837	30	208.
	Model F12-4H	30	449.
Ceiling Speaker			
Wide Angle Trumpet	Atlas		
	Model APC-30T 120 x 60 deg. disper.	30	143.
Flush Ceiling Speaker	Lowell		
	8" speaker mtd. in round back box and grille	15	91.
	12" cone with tweeter & woofer (incl transf.)	15	230.
	8" speaker and tranf. less enclosure	15	33.
Wide Range Auditorium Speaker	Rauland Borg		
	Model MLS-3A 30W		587.
	Model MLS-5 70W		740.
Volume Controls	Soundolier - Flush Mounted		
	Model AT-10 10W auto tranf.		59.
	Model AT-35 35W auto tranf.		60.

The costs shown for these sound system reinforcing components consist of the published contractors' book prices and include outlet boxes where required. Included are the necessary installation labor for equipment and a certain amount for supervision, final checkout, and training of operating personnel.

Type of Unit	Manufacturer	Amount
Microphone Receptacles	Switchcraft #G3MS	$ 57.
	Soundolier #MRB-2-14N	112.
Mixers	Shure #M-68	
	For 4 microphones and 1 auxiliary input- high & low imped. output	300.
P.A. Amplifiers-50 to 15,000Hz 4 Mike/Auxiliary Inputs	Rauland	
	#1402 20 watt	309.
	#1406 60 watt	575.
	#1410 100 watt	688.

The costs shown for these sound system reinforcing components consist of the published contractors' book prices and include back boxes as required. Included are the necessary installation labor for the equipment and a certain amount for supervision, final checkout, and training of operating personnel.

I-26

SOUND SYSTEM COMPONENTS

COST OF SOUND-REINFORCING SYSTEM COMPONENTS

Type of Unit	Manufacturer	Amount
Lavaliere Microphone	Shure #510S	
	For neck suspension w/30' cord	$206.
Microphone	Shure #519B Dynamic/omni directional with 20' cord	129.
	Shure #545SD Cardioid/directional with 15' cord	167.
	Turner #751 - Desk Mounted Paging microphone with 20' cord	94.

The costs shown for these sound system reinforcing components consist of the published contractors' book prices. Included are the necessary installation labor for the equipment and a certain amount for supervision, final checkout, and training of operating personnel.

Type of Unit	Manufacturer	Amount
Antenna Kit 	Blonder-Tongue AM/FM	$388.
AM/FM Tuner	Rauland Model SRX163	502.
Wires and Cables	West Penn #222 - one unshielded pair #300 - three shielded conductor and drain in a common jacket #292 - shielded 20 gage micro- phone cable with braid shield	.59 .69 .69

I-28

The costs shown for these sound system reinforcing components consist of the published contractors' book prices. Included are the necessary installation labor for the equipment and a certain amount for supervision, final checkout, and training of operating personnel. All cable is assumed to be pulled-in conduit.

COST OF HIGH-TEMPERATURE SIGNALING & POWER-LIMITED CABLE

	Gage	Number of Pairs or Cond	Cost Installed Per Foot
Non-Shielded	22 sol	1 pr	$.41
	18 str	1 pr	.63
	18 str	3/c	.73
	16 sol	1 pr	.72
Overall Foil Shield	24 sol*	2 pr	.58
	24 sol	25 pr	2.24
	22 sol	1 pr	.58
	22 sol	2 pr	.66
	22 sol	3 pr	.77
	22 sol	4 pr	.90

*braided shield with drain wire

The costs shown for high-temperature signaling and power-limited cables consist of the published contractors' book prices for Teflon-jacketed cable approved for use in environmental air plenums. Included is the labor for installation without conduit in accessible ceiling cavities.

Type of Unit	Description	Amount
Room Station for Toilet and Bedroom 	Jeron #2320	$38.
Apartment Door Release 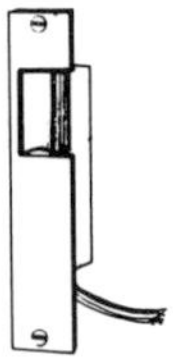	Jeron #8109	92.
Corridor Light 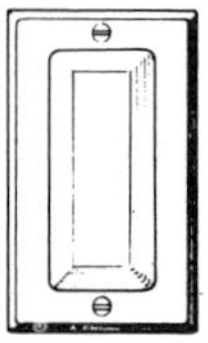	Jeron #8501	38.
Corridor Bell 6" dia. 	Jeron #8602	76.
Transformer 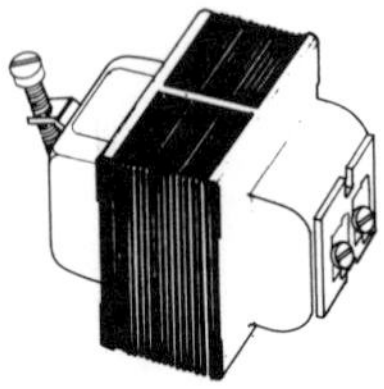	Jeron #8010	49.

I-30

The equipment shown here is of a design suggested by the Department of Housing and Urban Development, is frequently used in housing for the elderly, and is of an economical style. Costs shown include a back box where required and labor for installation of box and equipment.

EMERGENCY CALL SYSTEM FOR ELDERLY HOUSING

Type of Unit	Description	Amount
Photoelectic Smoke Detector 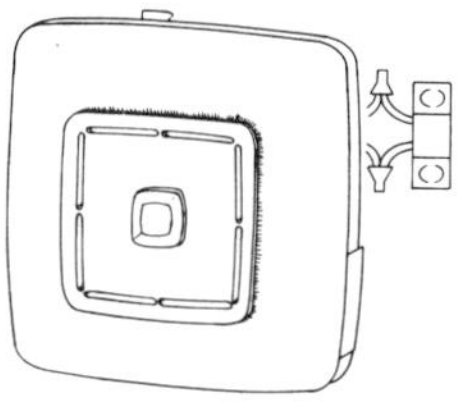	Gentex #710TCC	$108.
Remote Indicator for Office 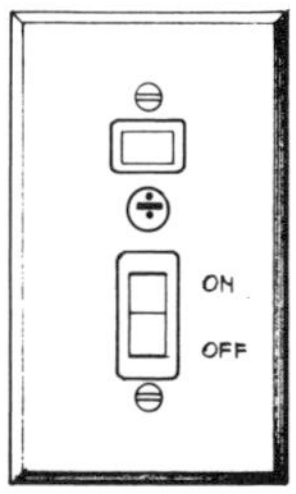	Jeron #8827	99.
Emergency Phome for Elevator to Manager's Office 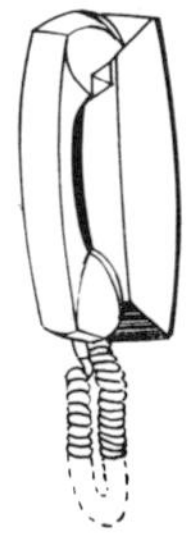	Jeron #2037	86.
Telephone Power Sypply 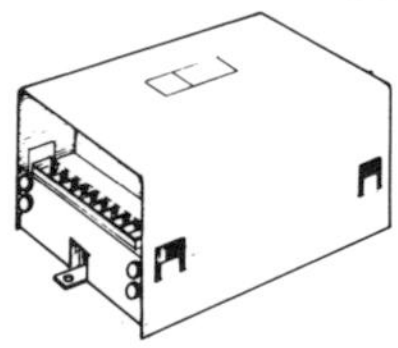	Jeron	132.
Master Annunciator Buzzer ans Silencing Switch Similar to Master Unit Button Plate on I33	Jeron #8701 10 Button 20 Button 30 Button 40 Button	301. 453. 597. 741.

The equipment shown here is of a design suggested by the Department of Housing and Urban Development, is frequently used in housing for the elderly, and is of an economical style. Costs shown include a back box where required and labor for installation of box and equipment.

I-31

Type of Unit	Description	Amount
Apartment Speaker Unit	Jeron	
	#2001 (Push Button Talk Control)	$48.
	#2005 (Voice Actuated Control)	46.
Door Lock Release	Jeron #8109	
		92.
Power Supply	Jeron #8010	
	40 VA, 24VAC	49.
Amplifier	Jeron	
	#5010 (for P.B. talk sys)	196.
	#5020 (dual lobby application)	288.
	#5050 (for voice actuated sys)	406.

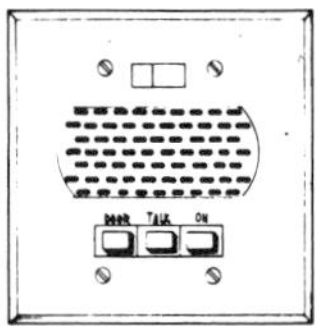

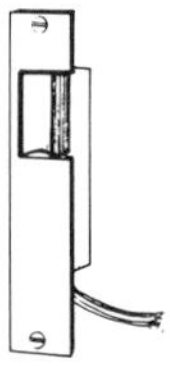

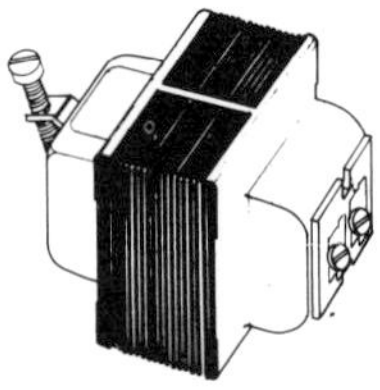

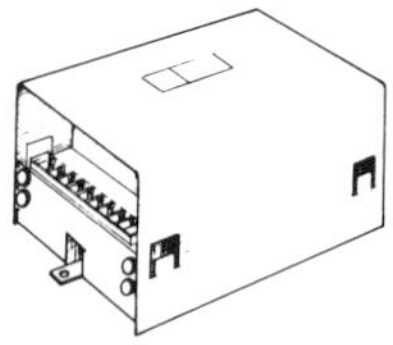

I-32

The equipment shown here is of a design suggested by the Department of Housing and Urban Development, is frequently used in housing for the elderly, and is of an economical style. Costs shown include a back box where required and labor for installation of box and equipment.

Type of Unit	Description	Amount
Directory 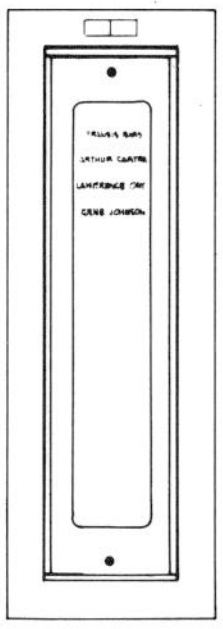	Jeron #3018 Up to 40 names	$ 94.
Combination Master Unit And Entrance Speaker With Plastic Buttons 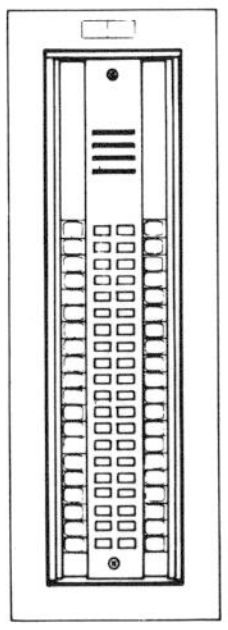	Jeron #3003 10 Button 20 Button 30 Button 40 Button For metal buttons, Add	 $236.00 355.00 468.00 578.00 2.20
Master Unit Button Plate (No Speaker) 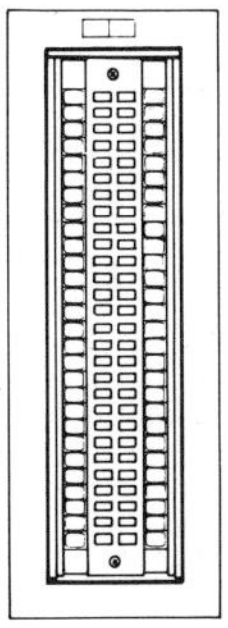	Jeron #3005 10 Button 20 Button 30 Button 40 Button 50 Button	 223.00 344.00 458.00 571.00 684.00

The equipment shown here is of a design suggested by the Department of Housing and Urban Development, is frequently used in housing for the elderly, and is of an economical style. Costs shown include a back box where required and labor for installation of box and equipment.

Type of Unit	Manufacturer	Amount
Roof Antenna	Commercial Grade Broad Band Winegard CH-7082	$533.
	Commercial Grade Single Channel - Yagi Type Winegard Add for additional channel	454. 198.
	Residential Grade 9 Element 300 ohm Broadband Winegard CH-7080	245.
Amplifier	Broad Band Low Gain 22dB Winegard DA-805 Med. Gain 33dB Winegard DA-830 High Gain 55dB Winegard DA-8245 Single Channel 42dB Winegard DX-02XX 60dB Winegard DX-03XX	245. 373. 530. 448. 573.
Signal Splitter	Two Way Winegard TV1030 Four Way Winegard TV1035	24. 31.

Costs consist of published contractors' book price for materials including anchoring and fastening materials. Antennas are mounted on an 8-foot-high mast guyed in three directions. Amplifiers are mounted on plywood board and include input and output coaxial cable connections to next component.

COST OF TV MASTER ANTENNA SYSTEM

Type of Unit	Manufacturer	Amount
Wall Outlet	Fixed Loss - Winegard SCT77	$34.
	Variable Loss - Winegard VTF77	36.
Fixed Attenuating Pad	1, 3, 6, 10, and 20bB Winegard IP series	26.
Receiver Matching Transformer (Balun)	Winegard TV-3005 UHF/VHF	23.
Coax Cable Terminator	Winegard MSJ-6 (mounts on end outlet with transformer)	25.
Cable	RG59 (6db/100ft. CH13) RG6 (3db/100ft. CH13) RG11 (2.1db/100ft. CH13)	.94 1.14 1.57

Costs consist of published contractors' book price for materials and include outlet box for wall outlet. Coaxial cable is costed as pulled-in conduit. Refer to Section F for cost of conduit.

I-35

Electric Heating and Controls

CEILING OR SUSPENDED TYPE

Forced-Air Unit Heaters—Horizontal or Vertical   J-1
Infrared Radiant Heaters   J-2
Miscellaneous Ceiling Heaters   J-10

WALL TYPE

Residential Baseboard Heaters   J-3
Commercial Baseboard Heaters   J-4
Commercial Sill-Height Convectors   J-5, J-6, & J-7
Commercial Cabinet Convectors   J-8
Recessed Wall Heaters with Fan   J-9
Miscellaneous Wall Heaters   J-10

FREEZE PROTECTION

Gutter- and Downspout-Heating Cable   J-11
Pipe-Heating Cable   J-11

FLOOR HEATING

Heating Cable Mats   J-12
Floor Drop-in Heater—Forced Air   J-10

SNOW MELTING

Snow-Melting Cable Mats   J-13
Jacketed Mineral-Insulated Snow-Melting Cable   J-14

CONTROLS

Line-Voltage Thermostats   J-15
Low-Voltage Thermostats   J-16
Miscellaneous Controllers   J-16

J

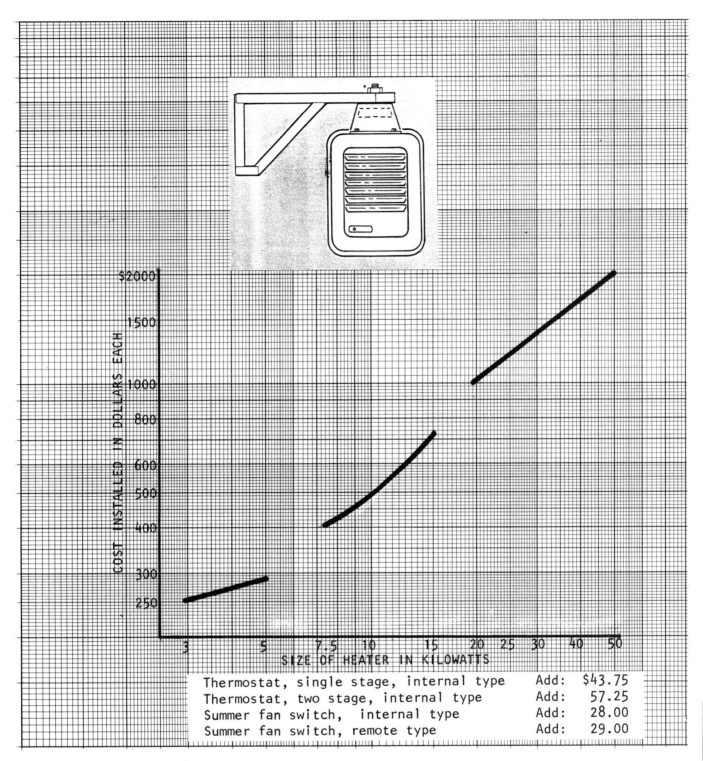

COST INSTALLED IN DOLLARS EACH

SIZE OF HEATER IN KILOWATTS

Thermostat, single stage, internal type	Add:	$43.75
Thermostat, two stage, internal type	Add:	57.25
Summer fan switch, internal type	Add:	28.00
Summer fan switch, remote type	Add:	29.00

*The costs shown for these units consist of the published contractors'
book prices for the units shown and the labor for installation. The
commercial-quality unit is manufactured by Emerson-Chromalox and
is type MUH. Included are a universal wall and a ceiling-mounted
bracket. No branch circuit wiring or controls are included.*

J

J-1

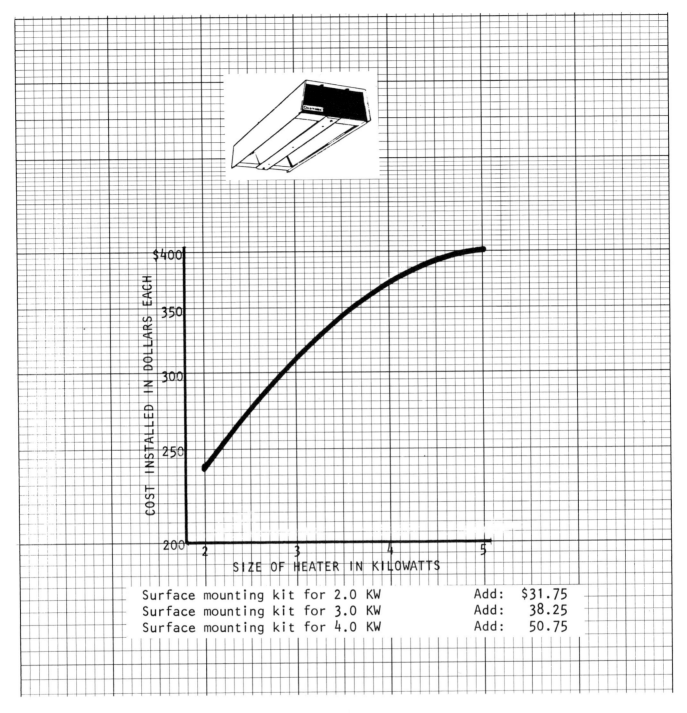

Surface mounting kit for 2.0 KW   Add: $31.75
Surface mounting kit for 3.0 KW   Add: 38.25
Surface mounting kit for 4.0 KW   Add: 50.75

*The costs shown for these units consist of the published contractors'
book prices for the units shown and the labor for installation. These
commercial units, sometimes called "people heaters," are manufactured
by Emerson-Chromalox and are its type RDO. They are UL-labeled for
either indoor or outdoor use with the bottom of the unit installed no
lower than 84 inches above the floor. Heating elements are included,
but no branch circuit wiring or controls.*

J-2

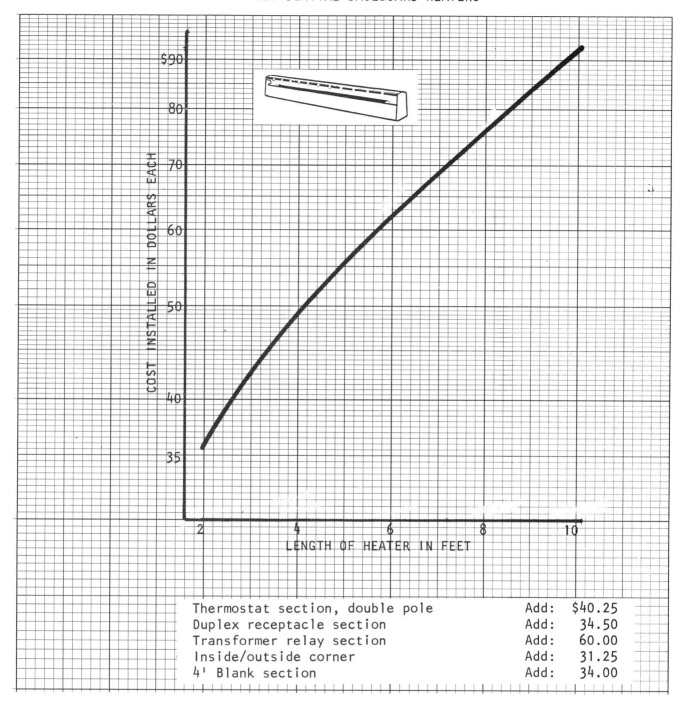

Thermostat section, double pole	Add:	$40.25
Duplex receptacle section	Add:	34.50
Transformer relay section	Add:	60.00
Inside/outside corner	Add:	31.25
4' Blank section	Add:	34.00

*The costs shown for these units consist of the published contractors'
book prices for the units shown and the labor for installation. These
residential units are manufactured by Emerson-Chromalox, are called
Space Board units, and are UL-listed. They are rated at 250 watts per
linear foot. No branch circuit wiring or controls are included.*

J

J-3

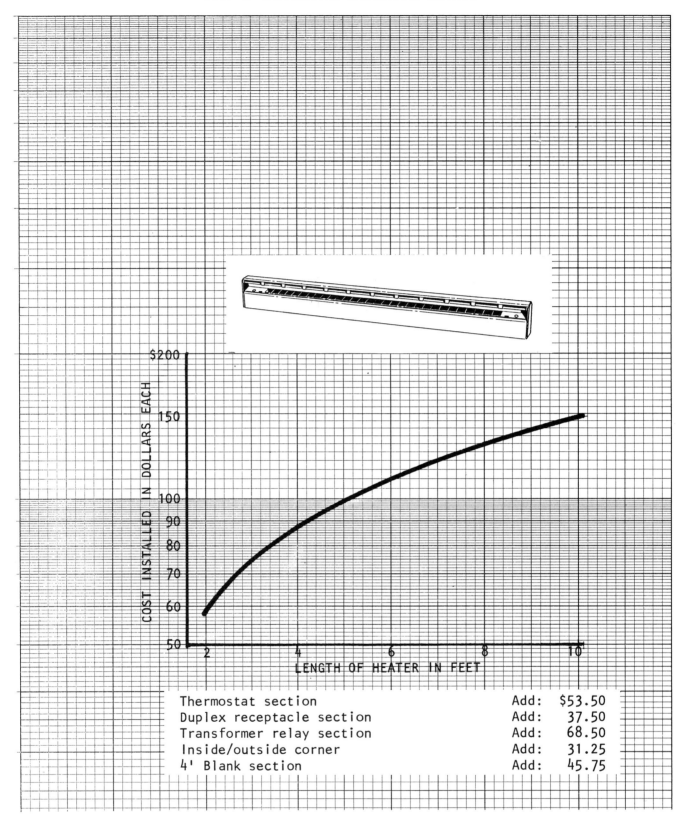

COST INSTALLED IN DOLLARS EACH

$200
150
100
90
80
70
60
50

2   4   6   8   10

LENGTH OF HEATER IN FEET

Thermostat section	Add:	$53.50
Duplex receptacle section	Add:	37.50
Transformer relay section	Add:	68.50
Inside/outside corner	Add:	31.25
4' Blank section	Add:	45.75

J-4

The costs shown for these units consist of the published contractors'
book prices for the units shown and the labor for installation. These
commercial-grade units are manufactured by Emerson-Chromalox. They
are the BB-C Series, UL-listed, and are rated at 250 watts per linear foot.
No branch circuit wiring or controls are included.

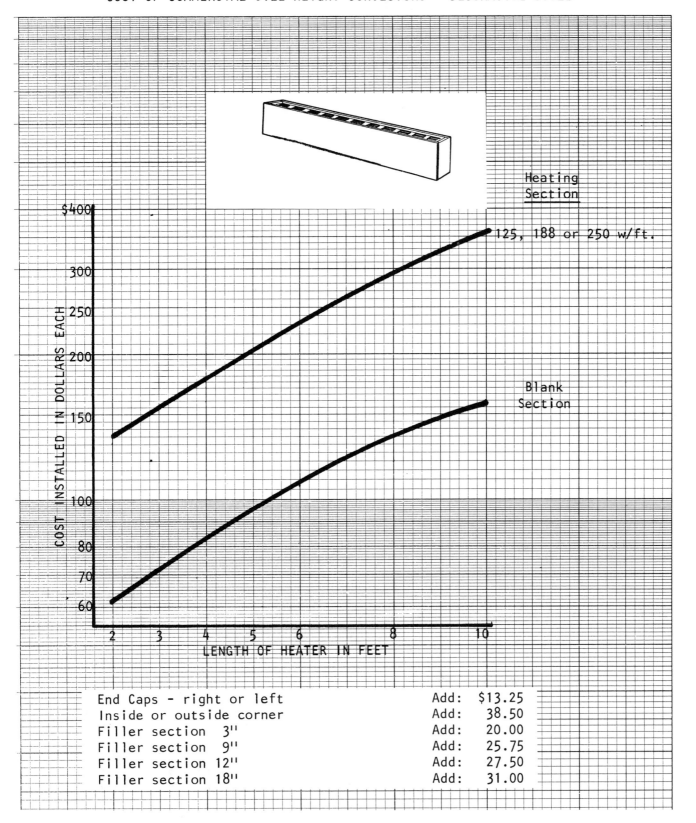

Heating
Section

125, 188 or 250 w/ft.

Blank
Section

COST INSTALLED IN DOLLARS EACH

$400
300
250
200
150
100
80
70
60

LENGTH OF HEATER IN FEET

2   3   4   5   6   8   10

End Caps - right or left	Add:	$13.25
Inside or outside corner	Add:	38.50
Filler section 3"	Add:	20.00
Filler section 9"	Add:	25.75
Filler section 12"	Add:	27.50
Filler section 18"	Add:	31.00

J

J-5

*The costs shown consist of the published contractors' book prices and
the labor for installation. These units are manufactured by Emerson-
Chromalox in the DSH Series in the wattages shown on the graph.
Blank sections are available to maintain architectural appearance where
heating sections are not required. No branch circuit wiring or controls
are included.*

5½" High X 3" Deep COMMERCIAL SILL-HEIGHT CONVECTORS

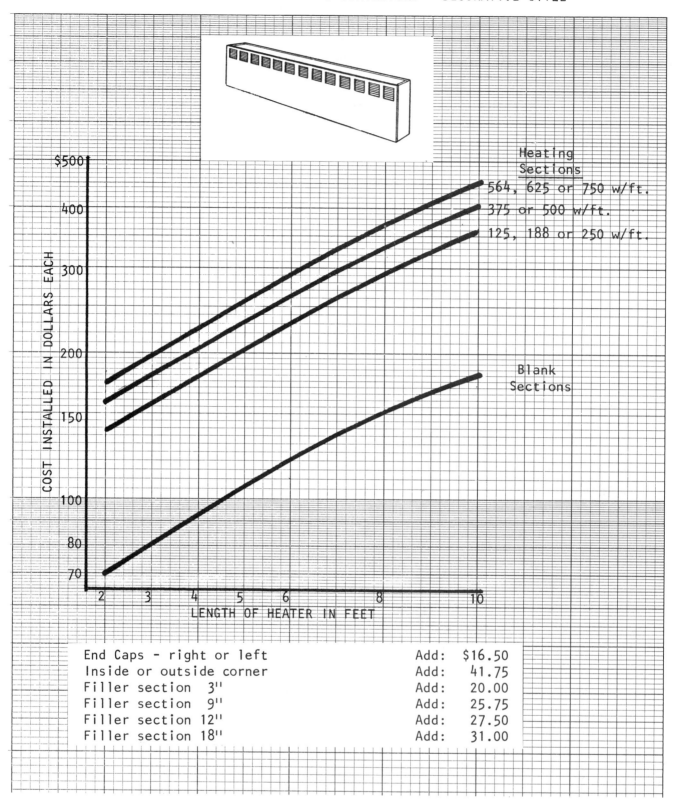

The costs shown consist of the published contractors' book prices and the labor for installation. These 7-inch-high units are manufactured by Emerson-Chromalox in the DSH Series in the wattages shown on the graph. Blank sections are available to maintain architectural appearance where heating sections are not required. No branch circuit wiring or controls are included.

J-6

7" High x 5" Deep COMMERCIAL SILL-HEIGHT CONVECTORS

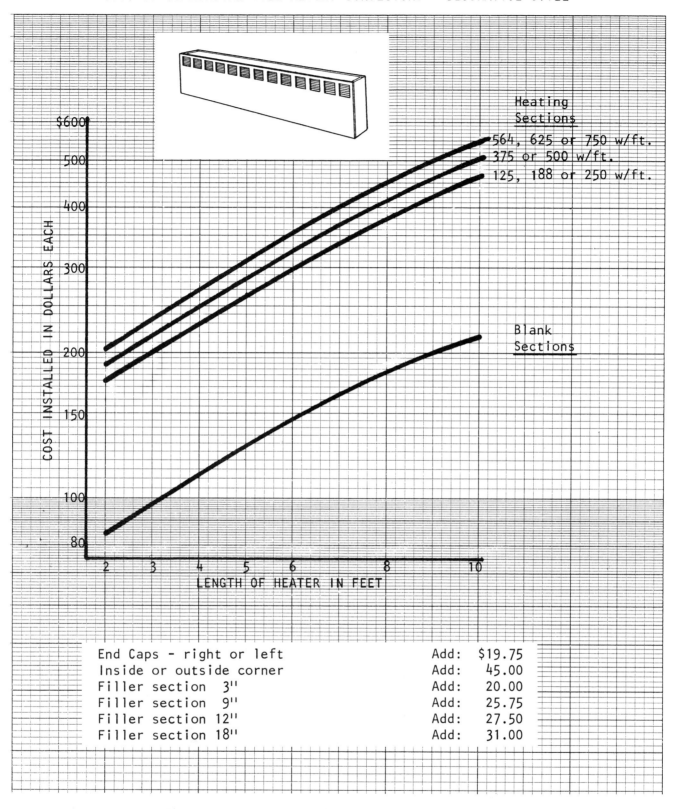

Heating Sections

564, 625 or 750 w/ft.
375 or 500 w/ft.
125, 188 or 250 w/ft.

Blank Sections

COST INSTALLED IN DOLLARS EACH

LENGTH OF HEATER IN FEET

	Add:
End Caps - right or left	$19.75
Inside or outside corner	45.00
Filler section 3"	20.00
Filler section 9"	25.75
Filler section 12"	27.50
Filler section 18"	31.00

J

The costs shown consist of the published contractors' book prices and the labor for installation. These 14-inch-high units are manufactured by Emerson-Chromalox in the DSH Series in the wattage densities shown on the graph. Blank sections are available to maintain architectural appearance where heating sections are not required. No branch circuit wiring or controls are included.

14" High x 5" Deep COMMERCIAL SILL-HEIGHT CONVECTORS

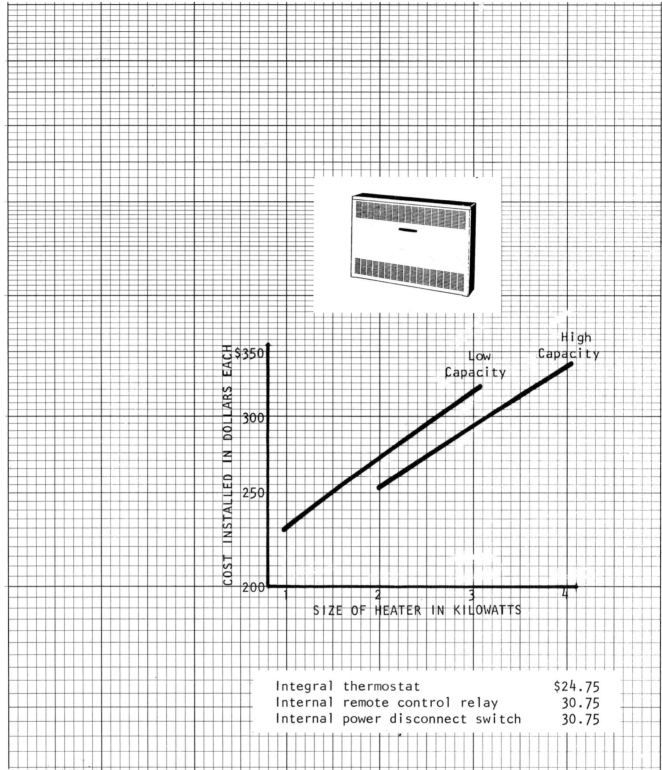

Integral thermostat $24.75
Internal remote control relay   30.75
Internal power disconnect switch   30.75

*The costs shown for these units consist of the published contractors'
book prices for the units shown and the labor for installation. These
commercial units are the standard type of the KSF and KSR Series as
manufactured by Emerson-Chromalox. The KSF is for surface mounting,
and the KSR is for recessed mounting. These units have no fans. No
branch circuit wiring or controls are included in the prices.*

COST OF RECESSED WALL HEATERS WITH FAN

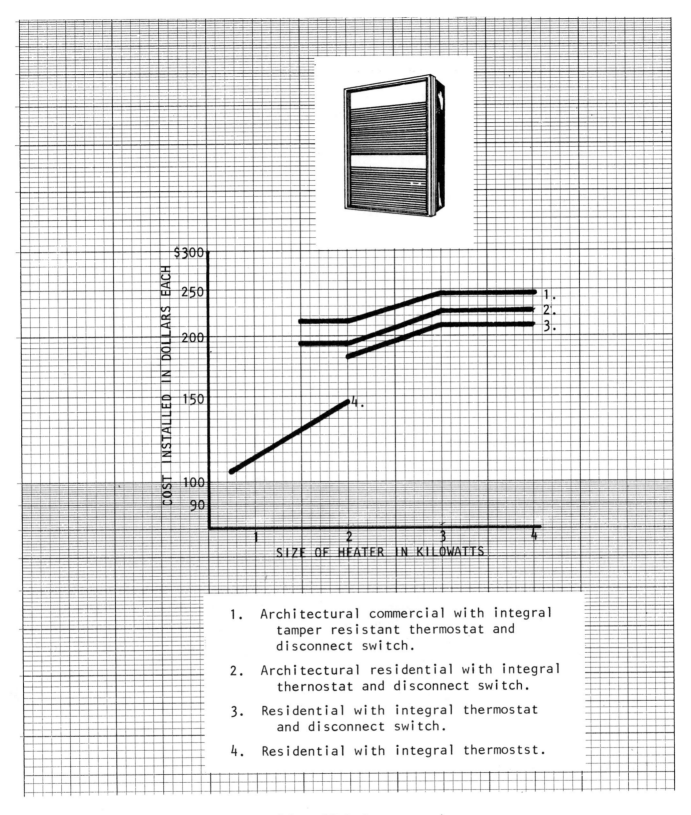

1. Architectural commercial with integral tamper resistant thermostat and disconnect switch.

2. Architectural residential with integral thermostat and disconnect switch.

3. Residential with integral thermostat and disconnect switch.

4. Residential with integral thermostst.

The costs shown for these units consist of the published contractors'
book prices and the labor for installation. These units are manufactured
by Emerson-Chromalox and are the AWH and FWH Series. No branch
circuit wiring or separate controls are included.

COST OF MISCELLANEOUS ELECTRIC HEATERS

Type of Unit		Manufacturer	Wattage	Installed Cost
	Recessed bathroom heater with built-in thermostat & fan.120v	Chromalox H-5022	1250	$90.25
	Fan forced ceiling heater. 120v	Chromalox #5014	1500	90.50
	Fan forced floor drop-in heater with comb-ination wattage at 120 volts. Kit for concrete slab mounting.	Chromalox FD-1415	375 750 1125 1500	178.75
	Fan forced kick space heater. Stat. kit for integral	Chromalox KSH 2000	1800	174.50 38.25
	Radiant ceiling pnl. With surface mtg kit With recessed mtg kit	Chromalox CP-754	750	127.25 161.25 170.00

The costs shown for these units consist of the contractors' book prices for the units shown. These units are considered to be installed but do not include any controls or branch circuit wiring. However, the wiring connection is included. The manufacturer and catalog numbers are as shown,

J-10

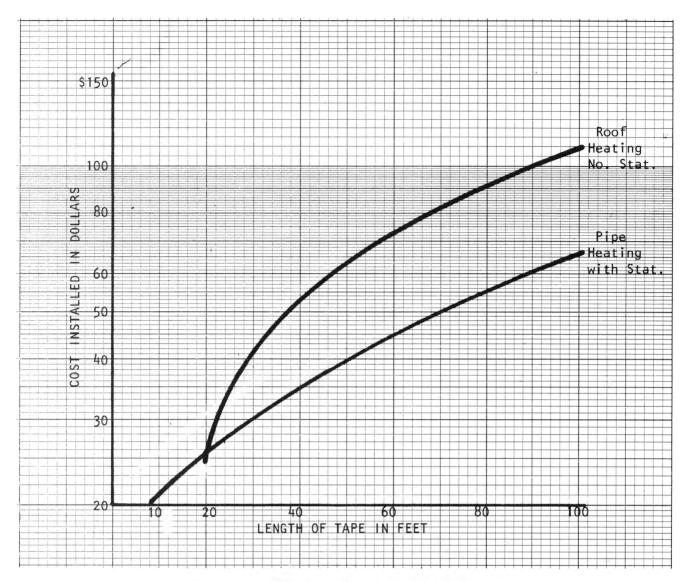

The costs of these units consist of the published contractors' book prices and the labor for installation. These tapes are manufactured by Emerson-Chromalox and are of the TBT and TWRK Series. Pipe tracing consumes an average of 1 1/2 feet of tape per foot of pipe protected. Roof-heating tape is clipped in place with manufacturers' clips. Both types are 120-volt. The pipe-heating cable has a built-in thermostat and is UL-listed. It is rated at 6 watts per linear foot. The roof-heating cable has a 6-foot cord, plug, clips, and spacers and is rated at 8 watts per linear foot.

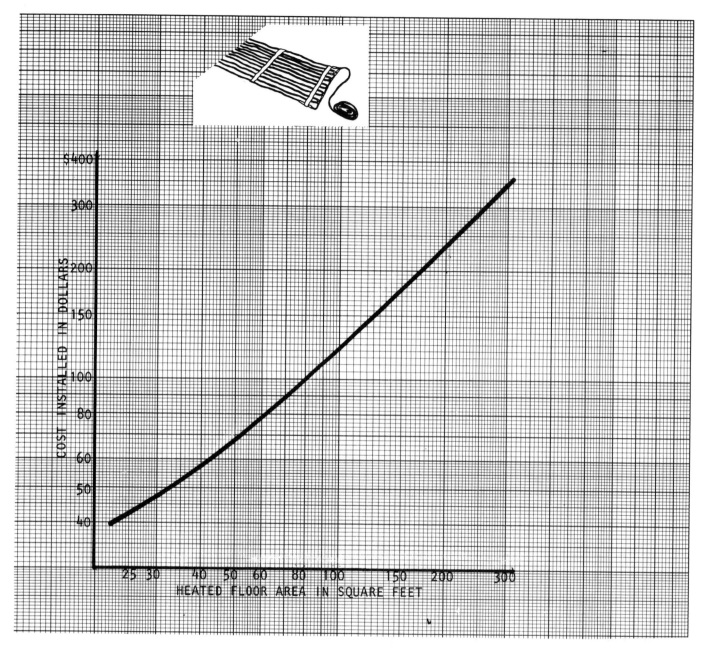

The costs shown for these units consist of the published contractors' book prices and the labor for installation. These mats are manufactured by Emerson-Chromalox and are the CTW Series rated at 20 watts per square foot and are for indoor use only. The heating wire is insulated with 90° PVC and bonded to spacer tapes. No branch circuit wiring or controls are included since these are influenced by design considerations. Mats are 16 inches wide.

J-12

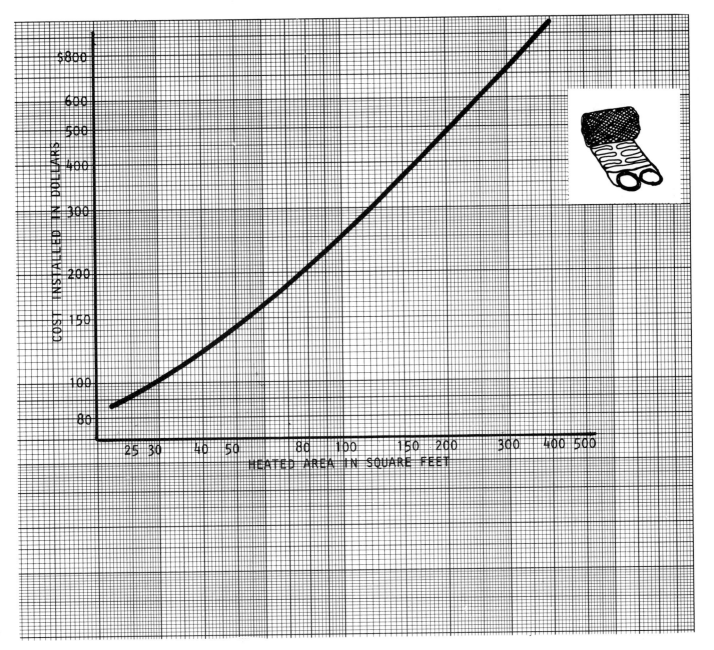

The costs shown consist of the published contractors' book prices
for the units shown and the labor for installation. This cable is
laid in place and is considered to be made of mats 18 inches wide
manufactured by Emerson-Chromalox in the TW Series at 48 watts
per square foot. It is of a grounded type and designed for outdoor
use. To prevent damage to the cable, the top 2 inches of fresh
asphalt should not exceed 350°F. Branch circuit wiring or controls
are not included since these are influenced by design considerations.

J

J-13

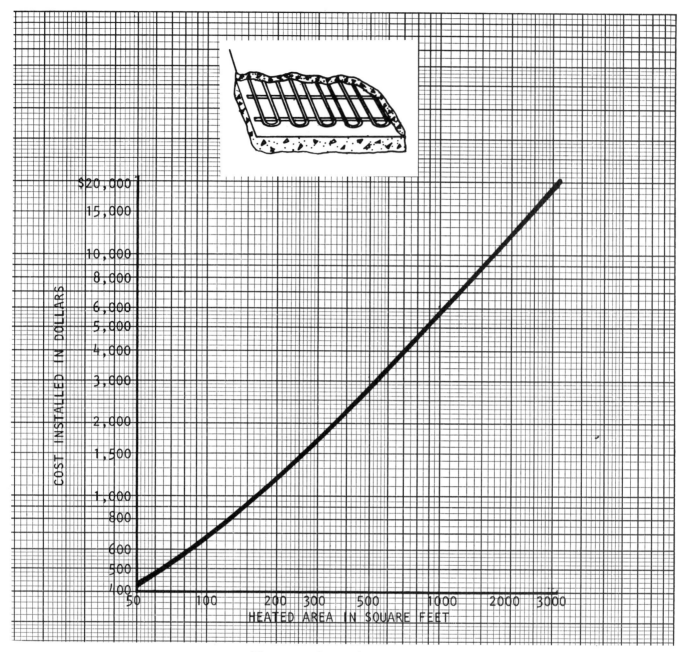

The costs shown for these cables consist of the contractors' published book prices for the cables required. The curve is based upon a wattage density of 45 watts per square foot. However, density can vary from 20 to 100 watts per square foot with an insignificant change in selling price. The cables are based upon Pyrotenax MI electric heating cable, single-conductor, with 7 feet of cold lead of the proper size on each end, factory-installed. All cable on this graph is 600-volt-rated. The graph further includes 4- by 4-inch by #8 welded wire mesh and the fastening of the heating wire to it on a concrete base course for 2 to 2 1/2 inches of finish pour for the heating surface. The mesh will help keep the heating cables in place during the concrete pour.

These costs do not include controls or branch circuit wiring to the heating cables since these requirements are a function of the design. Labor is included for the heating cable and cold leads only.

	Description	Manufacturer	Installed Cost
	Single stage - snap action DPST. Handles 5 KW @ 240v	Chromalox T-200	$52.25
	Single stage - tamperproof SPST. Handles 5 KW @ 240v	Chromalox MHT-4051E	94.00
	Single stage - heavy duty Snap-action DPST. Handles 5 KW @ 240v	Chromalox WR-80	94.50
	Two stage - modulating Handles 5 KW @ 240v per stage. 35 - 90°	Chromalox M-62	68.75
	Three stage - for pilot duty. 120 VA max. @ 120v per stage. 40 - 80°	Chromalox T-42M	156.75
	Single stage with 10' Capillary for floor heating control. Handles 22 amps @ 240v	Chromalox TWT-70D	99.75

J

The costs shown for these units consist of the published contractors' book prices and the labor for installation. They also include a recessed junction box with conduit terminals for mounting each thermostat.

Generally, line-voltage units do not have quite as narrow a range of control as units of the piloted type. No branch circuit wiring is included.

LOW VOLTAGE

Description		Manufacturer	Installed Cost
	Single Stage - SPST with Adjustable Heat Anticipator 30 VAC 50⁰ - 90⁰	Chromalox WR-1E30	$50.50

MISCELLANEOUS CONTROLLERS

	Description	Manufacturer	Installed Cost
	Silent Transformer - Relay Low Voltage Single Level Temperature SPST 5200 W @ 208V 6000 W @ 240V	Chromalox WR-24A0IG-8	45.00
	Silent Transformer - Relay Low Voltage Two Stage 6000 W @ 240/Stage	Chromalox **WR-24A06G-3**	69.75
	Adjustable Motorized Input Controller (for cycling infrared heaters)	Chromalox VCS501A-30	134.75
	Snow Detector Assembly and Control Box	Nelson SNO-41	851.00

J-16

The costs shown for these units consist of the published contractors' book prices for the units shown. These piloted ON-OFF control thermostats are identified by the manufacturer and catalog number All include a recessed junction box with conduit terminals for mounting the thermostat. Piloted units generally have a closer temperature control range than line-voltage units. They control through relays of various types. Installation labor is included.

Snow Detector with Sensor and Control Box
The snow detector consists of a control box and sensor units. The moisture sensor is embedded in a heated slab; the slab thermostat, in an unheated slab. The air thermostat is the remote-bulb type. There must be conduit runs from the sensor to the control unit, and these must be figured separately.

LOW-VOLTAGE THERMOSTATS AND MISCELLANEOUS CONTROLLERS

Power Distribution Above 600 Volts

OVERHEAD

Wood Poles, 30–45 feet   K-1
    Pole Delivery Cost   K-2
Poletop Structures   K-3
Pole Supports   K-4
Pole Accessories   K-5
Bare ACSR Primary Conductors   K-6

UNDERGROUND

15 kV—URD—Aluminum Cable   K-7
Cutouts, Arresters, and Terminations   K-8

COST OF DISTRIBUTION POWER POLES

INSTALLED COST IN DOLLARS			
Western Fir & Lodgepole Pine Poles			
Pole Length	Class of Pole		
	3	4	5	6
30	----	$257.00	$232.00	$210.00
35	----	310.00	280.00	257.00
40	$401.00	362.00	330.00	----
45	467.00	424.00	388.00	----

Class of pole definition			
Pole Length in feet	Class of Western Fir		
	3	4	5	6
	Minimum top circumference in inches		
	23	21	19	17
	Minimum circumference at 6' from butt in inches		
30	32.5	30.0	28.0	26.0
35	35.0	32.0	30.0	27.5
40	37.0	34.0	31.5	29.0
45	38.5	36.0	33.0	30.5

The poles referred to here are distributed by Koppers Corporation, Inc., which has a plant in Denver, Colo. They are fully pressure-treated with pentachlorophenol.

The prices shown include the cost of the poles in Denver (see page K-2 for the cost of trucking), clearing the site of the pole location in normal soil, digging the hole to the proper depth, setting the pole with proper equipment, and backfilling and tamping the earth around the pole after alignment. The associated labor is included.

K

K-1

Since the cost of pole delivery is somewhat dependent upon pole length, weight, and distance, the assumption has been made that the project involves one to five poles (delivery cost is the same for one or for five), not exceeding 30,000 pounds or longer than 45 feet (heavier and longer loads involve an additional rate). Poles 50 feet and longer are classified as transmission rather than distribution poles.

These costs represent "loaded distance" with 1 hour for loading and 1 hour for unloading. At this time additional time at either end costs about $37 per hour in the Denver area. The costs shown are for hauling within the state. Out-of-state hauling costs $2.25 per loaded mile.

COST OF POLE DELIVERY

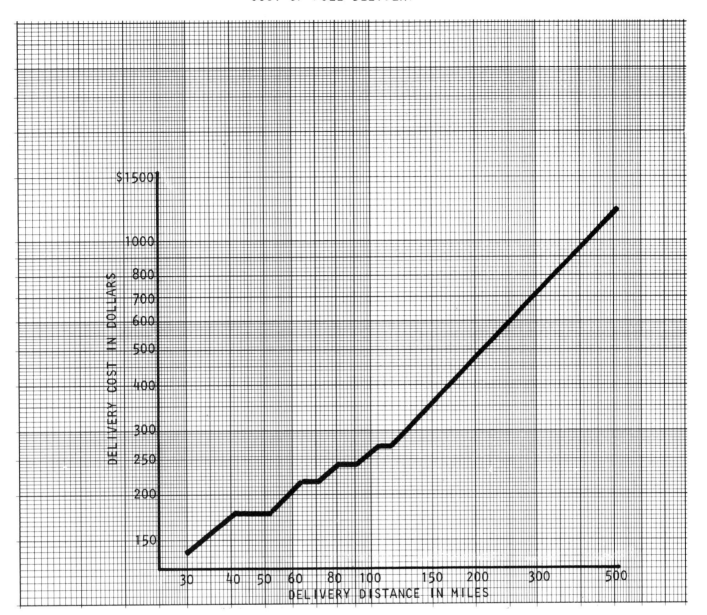

DELIVERY COST IN DOLLARS

DELIVERY DISTANCE IN MILES

K

K-2

The costs shown for these poletop structures consist of the published contractors' book prices for the various components making up a structure, the majority of which are Joslyn. The cost also includes the labor for the installation of the structure to the poletop at ground level before erecting the pole. The cost of the pole is not included.

COST OF POLETOP STRUCTURES

Description	Installed Cost
3-PHASE 4-WIRE 0-30° ANGLE — PLAN 5-30°	$220.00
PLAN 0-5°	133.00
Three Phase 4 Wire Single Dead End	289.00
Three Phase 4 Wire Double Dead End	465.00
Three Phase 4 Wire Tap at 0-5° Angle	394.00

Description	Installed Cost
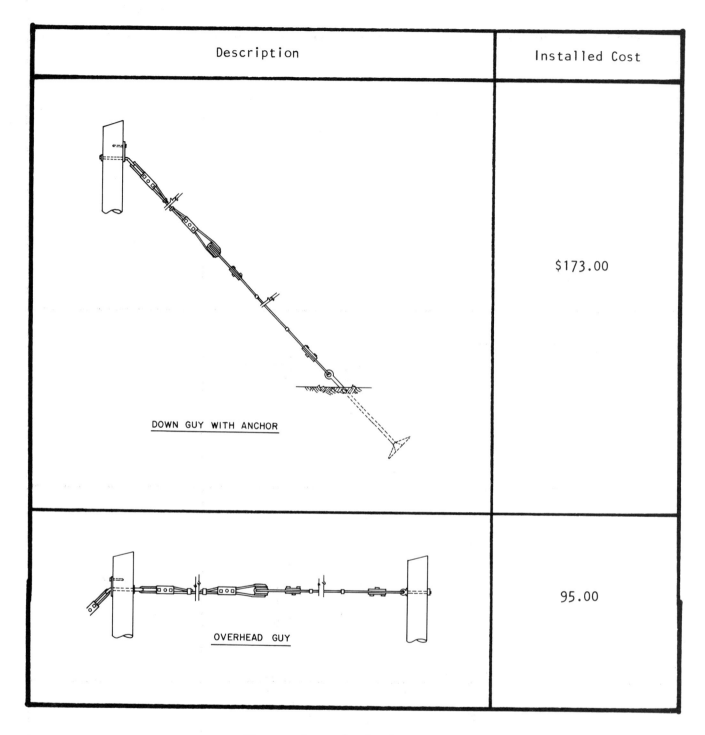 DOWN GUY WITH ANCHOR	$173.00
OVERHEAD GUY	95.00

The cost shown for the down guy consists of the published contractors' book prices for the various components shown. Included are an eight-way expanding anchor and 8-foot anchor rod. Also included is an 8-foot plastic guy guard. Labor includes drilling the hole by hand, placing the anchor, backfilling and tamping as required, and assembling and fastening the guy.

K-4

COST OF MISCELLANEOUS POLE-LINE ACCESSORIES

Description	Installed Cost
Aluminum Service Wedge Clamp	$22.00
3 Phase Transformer Cluster Mounting Bracket	90.00
Insulated Clevis and Thru Bolt	21.00
Messenger Hanger and Thru Bolt	20.00

The costs shown consist of the published contractors' book prices for the components shown and the labor for their installation.

K

K-5

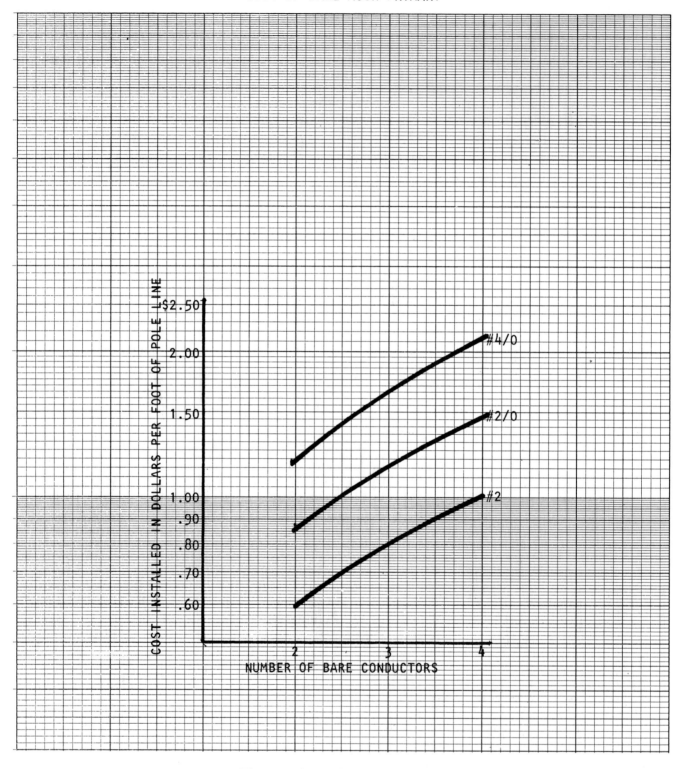

The cost shown for the ACSR bare conductor consists of the published contractors' book price and the labor for stringing, sagging, and tieing the conductors to the insulators. The cost shown represents the length of pole line and not the conductor footage.

K-6

COST OF PRIMARY UNDERGROUND DISTRIBUTION CABLE

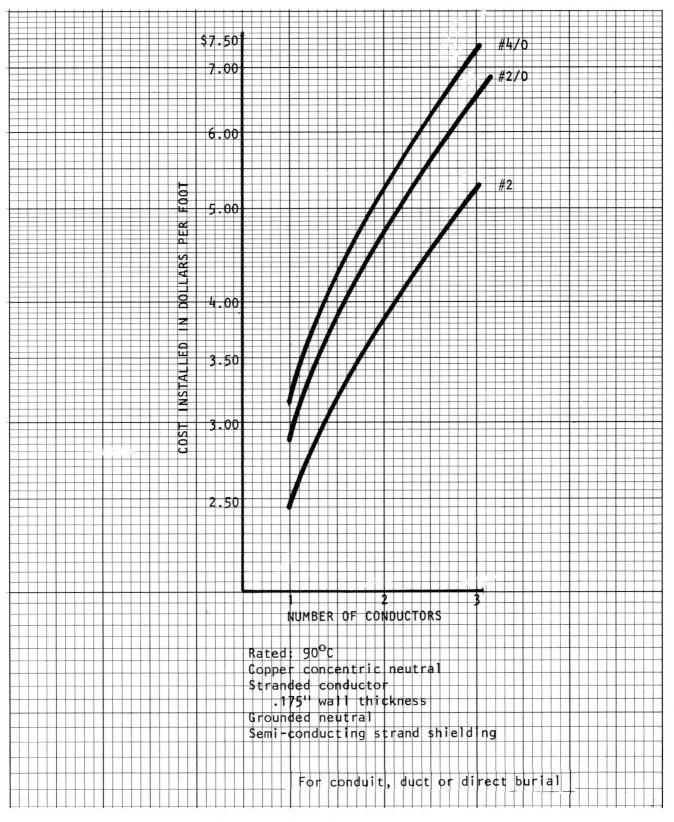

Rated: 90°C
Copper concentric neutral
Stranded conductor
.175" wall thickness
Grounded neutral
Semi-conducting strand shielding

For conduit, duct or direct burial

K

K-7

The cost for the cable consists of the contractors' price (Kaiser) for the size and type shown, purchased in 5000-foot quantity. Adjust price as required by length. The labor is also provided for installation in an open trench or pulled-in conduit.

15-KV CABLE - ALUMINUM - TYPE URD - CROSS-LINKED POLYETHYLENE

COST OF DISTRIBUTION CUTOUTS, ARRESTERS, AND TERMINATIONS

Type of Unit	Manufacturer	Rating	Installed Cost
Fused Cutouts	G. E. Load Break		
	7.8/13.8 KV	100 Amp.	$236.00
Lightning Arrester	G. E.		
	Metal Oxide	9 KV 27 KV	86.50 133.50
Exterior Pole Termination	Blackburn		
	Type P & PBB Bracket	#2 - #4/0 25 KV	85.50
Interior Termination Stress Relief Cone	Blackburn		
	Type SKD	15 KV	55.00
Primary Elbow Terminator	G. E.		
	With Test Point	L - N 8.3 KV #2, 1/0 & 4/0	80.50

The costs for these items consist of the published contractors' book prices for items illustrated.

Included: (material and labor)

1. Fastening of unit to crossarm as required
2. Making conductor termination as required

CUTOUTS, ARRESTERS, AND TERMINATORS

Miscellaneous

JUNCTION BOXES

Junction Boxes   L-1

TRENCHING AND BACKFILLING

Trenching and Backfilling   L-2

DRILLING HOLES

Drilling Holes in Masonry   L-3
Drilling Holes in Reinforced Concrete   L-4
Cutting Holes in Steel Panel   L-5

CHANNELING FOR CONDUIT

Channeling of Masonry and Reinforced Concrete   L-6

CUTTING PAVEMENT

Cutting Asphalt Pavement   L-7
Cutting Concrete Pavement   L-8

ANCHORS

Toggle Bolts, Wall Anchors, & Expansion Shields   L-9
Hammer-Drive, Stud, & Self-Drilling Anchors & Self-Tapping Screws   L-10

L

COST OF JUNCTION BOXES

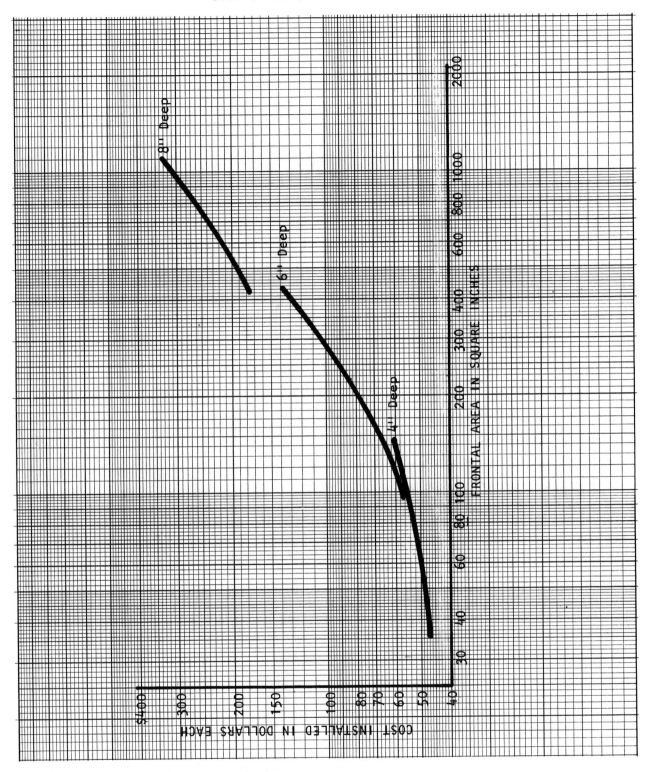

The costs shown for the junction boxes consist of the contractors'
published book prices. The boxes are assumed to be surface-mounted
on masonry. Included are the fastening devices. However, no conduit
terminals or holes are provided.

PULL OR JUNCTION BOXES

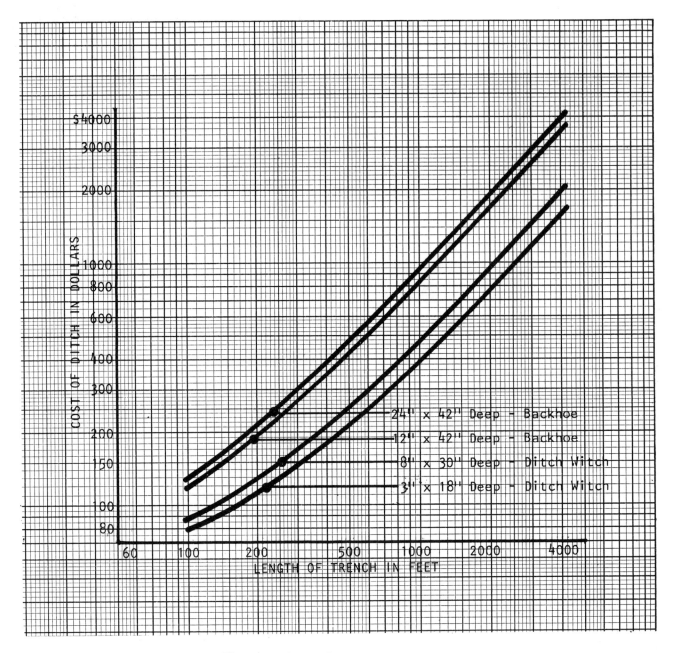

The costs shown for trenching and backfilling are based upon the size of trench shown on the graph. Normal, average dirt-digging conditions are assumed. The estimator must use job factors for more difficult digging conditions or ditch sizes different from those shown. Generally speaking, the cost is influenced by the volume of the dirt that has to be removed:

Backfill: Where no specific density except that which can be obtained by the equipment wheel rolling over the ditch is required, add 50% to the trenching cost. For backfill requiring 90 to 95% machine compaction, as you might find under parking areas or roads, double the trenching costs.

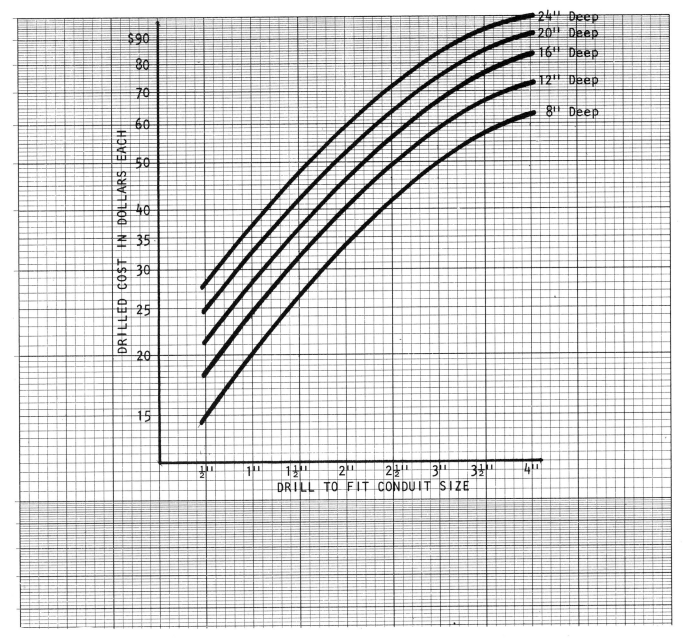

*The costs shown for drilling holes in masonry are based upon the use of
a diamond-core drill and include wear and tear on the drill.*

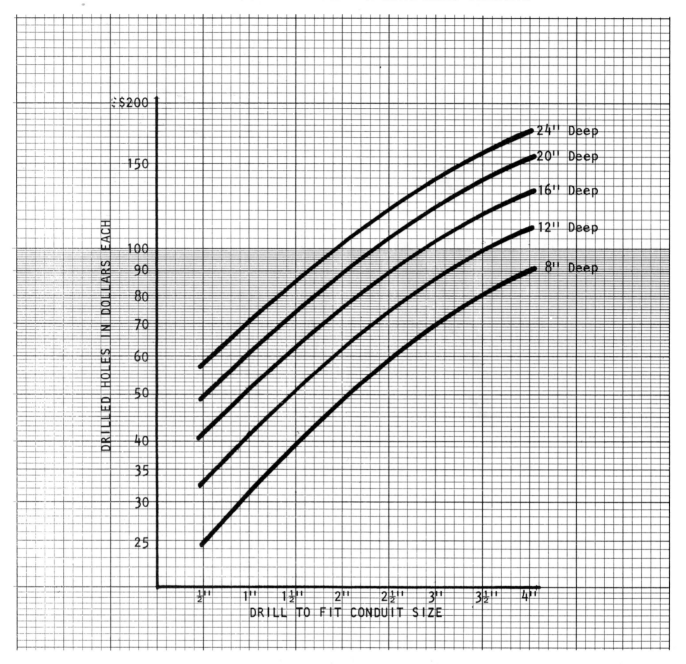

The costs shown for drilling holes in reinforced concrete are based upon the use of a diamond-core drill and include wear and tear on the drill.

DRILLING HOLES IN REINFORCED CONCRETE

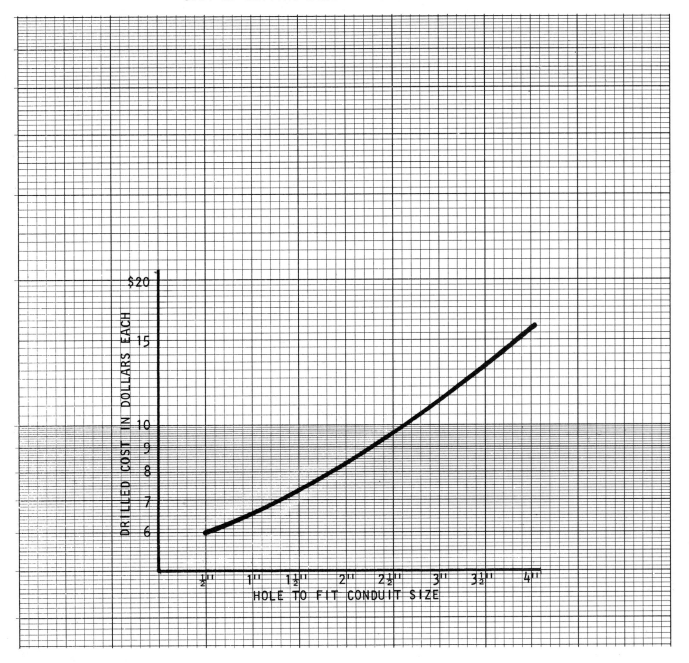

The costs shown here represent the costs of punching a hole in steel by using a hydraulic punch. If a hole saw is to be used, add 20%.

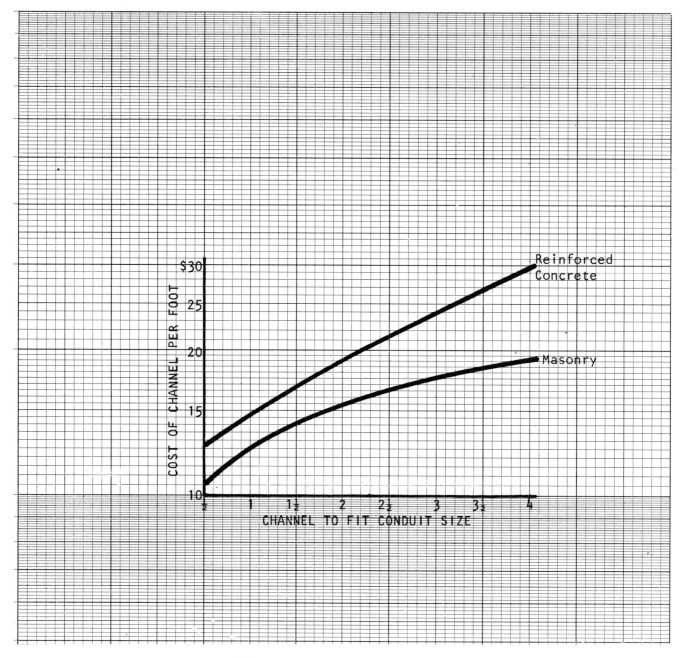

The costs shown for channeling in masonary or concrete are based upon two cuts with a power masonry saw, and an electric hammer for chipping out the trench is assumed to be available. The depth of cut is based upon about 5/8 inch of cover over the conduit. Dry-mix cement and labor for its installation are included in the costs shown. Also included are the wear and tear on the masonry blades.

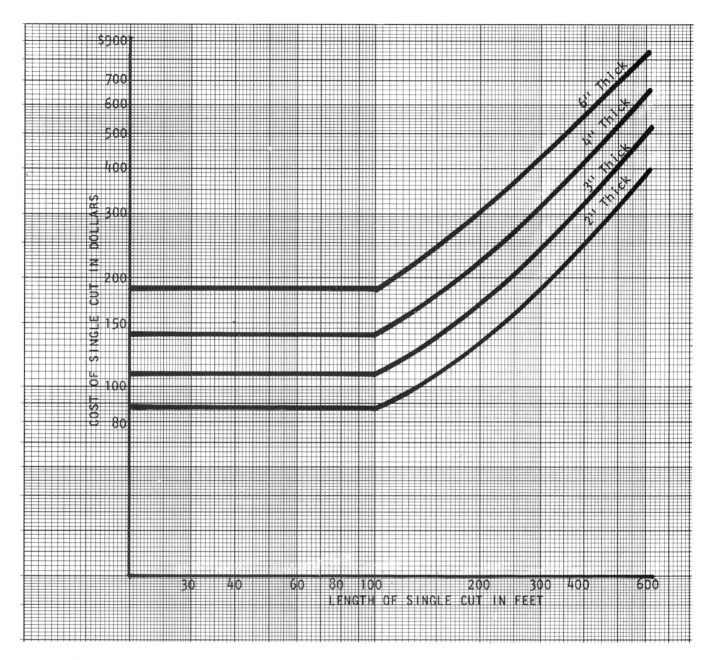

The costs shown for cutting asphalt pavement are based upon a single cut with a power masonry saw. The depth of cut as required to cut through the thickness of the asphalt is indicated on the graph. Bear in mind that you must add for handling debris and asphalt replacement as job conditions dictate.

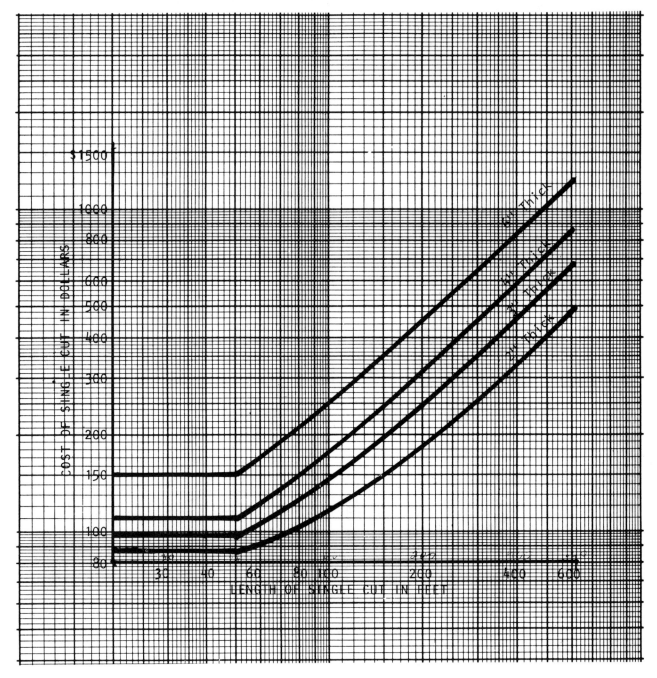

The costs shown for cutting concrete are based upon a single cut with a power masonry saw. Bear in mind that you must add for handling debris and concrete replacement as job conditions dicate.

COST OF WALL FASTENINGS

Description	Fastened To Drywall	Fastened To Hollow Masonry	Fastened To Concrete
Toggle Bolts — Rawl Toggle Bolt, GSA Specification FF-B-588-5306, Use in: Block, wallboard, plaster, hollow tile			
To 3/8" hole	$6.75	$7.25	
To 5/8" hole	8.50	9.00	
To 7/8" hole	10.75	11.25	
Hollow Wall Anchor — Rawly, GSA Specification FF-B-588-5340			
To 1/4" hole	7.75		
Over 1/4" hole	8.75		
Fiber or Plastic Anchor — Rawlplug, GSA Specification FF-S-325, Group IV, Type 2			
To #8 Screw	6.75	7.25	$7.75
To #14 Screw	8.50	9.00	9.50
To #20 Screw	10.25	10.75	11.25
Expansion Shield & Bolt — Rawl Single, GSA Specification FF-S-325, Group II, Type 2, Class 2, Style 1			
To 3/8" hole		$10.75	$11.25
To 5/8" hole		14.75	15.25
To 7/8" hole		19.75	20.25
Lead Anchor & Lag Screw — Rawl Lag-Shield, GSA Specification FF-S-325, Group II, Type 1, Classes 1 and 2			
To 3/8" hole		8.50	9.00
To 5/8" hole		10.75	11.25
To 7/8" hole		14.75	15.25

*The cost shown for these anchors consist of the published contractors'
book price and the labor for installing on the surfaces indicated.*

TOGGLE BOLTS, WALL ANCHORS, FIBER ANCHORS, EXPANSION SHIELDS

Description	Fastened To Metal	Fastened To Hollow Masonry	Fastened To Concrete
Hammer Drive Anchor Rawl Zamac Nailin GSA Specification FF-S-325, Group V, Type 2, Class 3 To 1" deep To 2" deep			$13.50 16.75
Stud Anchor & Nut Rawl-Stud — GSA Specification FF-S-325, Group II, Type 4, Class 1 To 3/8" hole To 5/8" hole To 7/8" hole			9.00 12.00 20.25
Self-Drill Anchor & Bolt Rawl Saber-Tooth — GSA Specification FF-S-325, Group III, Type 1 To 3/8" hole To 5/8" hole To 7/8" hole			10.25 14.00 21.00
Self-Tapping Screws in metal Rawl TEKS SLOTTED HEX HEAD, RABOT-NEOPRENE WASHER, HEX HEAD, PHILLIPS PAN HEAD #8 x 1" #12 x 1"	$4.75 6.00		

The costs shown here include the fastening devices plus the labor to make the installation to the surfaces shown.

HAMMER-DRIVE, STUD, & SELF-DRILLING ANCHORS & SELF-TAPPING SCREWS

Appendix

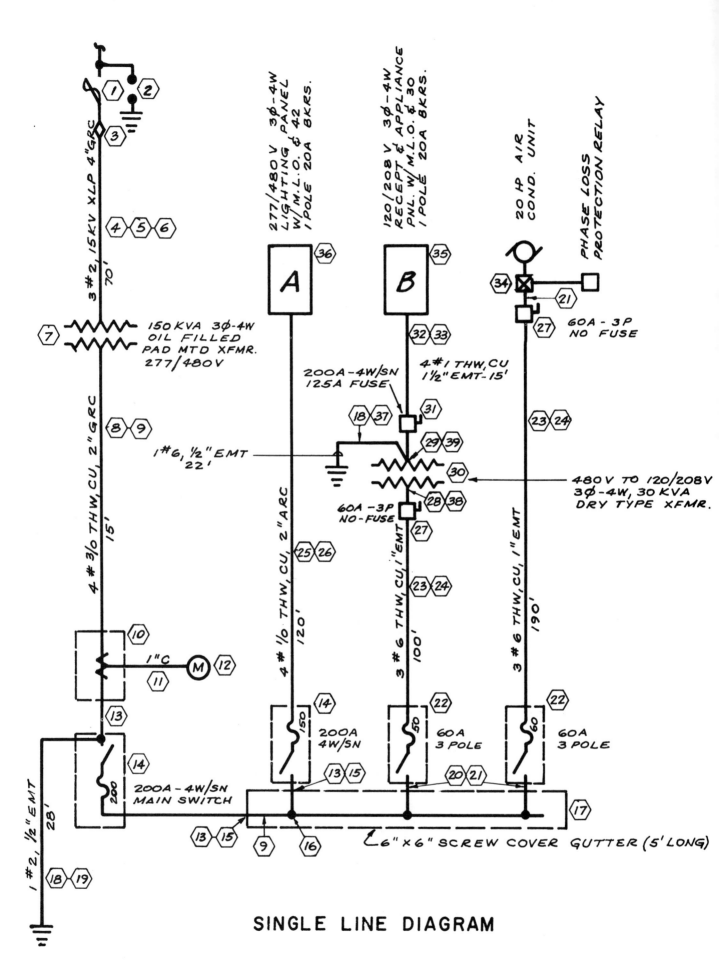

SINGLE LINE DIAGRAM

EXAMPLE NO. 1

REEP
PRICING SHEET

JOB _EXAMPLE #1_   SHEET NO. _1_

WORK _SINGLE LINE DIAGRAM_   OF _2_

PRICED BY _G. VoKo_   DATE _4/2/85_

DESCRIPTION	KEY	REMARKS or DEVIATIONS	REF PAGE	QUANTITY	UNIT PRICE		EXTENDED PRICE	
100a – 15KV FUSED CUTOUT	1	ON POLE	K-8	3	236	00	708	00
9KV LIGHTNING ARRESTER	2	"	K-8	3	86	50	259	50
EXTERIOR 15KV. CBL. TERMINATION	3	"	K-8	3	85	50	256	50
4" PVC COATED GRC	4	INCL 20' UP POLE	E-1	90'	19	50	1755	00
3#2 15KV XLPE CABLE	5	PULLED IN	K-7	105'	5	35	561	75
12"W. x 42"D. TRENCH	6		L-2	70'	—		120	00
150 KVA 3φ PAD MTD. XFMR	7	INCL PAD	H-9	1	—		5500	00
2" GRC	8		E-1	15'	4	10	61	50
4# 3/0 THW, CU	9		E-8	24'	7	30	175	20
200a – 4W INDOOR C.T. CBT.	10		C-2	1	—		570	00
1" GRC	11		E-1	15'	2	30	34	50
OUTDOOR METER HOUSING	12		—	1	—		50	00
2" NIPPLE ASSEMBLY	13		E-5	3	19	00	57	00
200a – 4W/SN FUSIBLE DISC. SW.	14	277/480	B-17	2	600	00	1200	00
CUT 2" HOLE IN GUTTER	15		L-5	2	8	50	17	00
FEEDER TAPS – #3/0 CONDUCTOR	16		E-17	10	26	25	262	50
6"x6" S.C. GUTTER	17		C-8	5'	—		83	00
1/2" EMT	18		F-5	50'	1	35	67	50
1#2 BARE COPPER GROUND	19		D-2	30'	2	00	60	00
CUT 1" HOLE IN GUTTER	20		L-5	2	6	50	13	00
1" NIPPLE ASSY.	21		E-5	3	10	75	32	25
60a – 3P FUSIBLE DISC. SWITCH	22	600V.	B-18	2	195	00	390	00
1" EMT	23		E-1	290'	1	80	522	00
3#6 THW, CU	24		E-8	290'	1	75	507	50

DESCRIPTION	QTY.	UNIT LABOR	EXTENDED LABOR HOURS	UNIT MAT'L.	EXTENDED MAT'L. $	
						13,263 70
						SHEET TOTAL (Unadjusted)

Total labor hours				
Labor rate per hour		× 16.00		
Total labor dollars				
Multiplier		× 1.97		
Total labor, overhead, & direct job expenses				
Total material cost				
Total gross cost				× 1.05 = —

REEP
PRICING SHEET

JOB __EXAMPLE #1__ SHEET NO. __2__
WORK __SINGLE LINE DIAGRAM__ OF __2__
PRICED BY __G. V. K.__ DATE __4/2/85__

DESCRIPTION	KEY	REMARKS or DEVIATIONS	REF PAGE	QUANTITY	UNIT PRICE		EXTENDED PRICE	
2" ALUM. RIGID CONDUIT	25		E-1	120'	4	10	492	00
4 #4/0 THW, CU	26		E-8	120'	8	70	1044	00
60a - 3 POLE NO FUSE DISC. SW.	27	600V.	B-19	2	110	00	220	00
1" CONDUIT TERMINAL (EMT)	28		E-2	1	—		7	10
2" CONDUIT TERMINAL (EMT)	29		E-2	1	—		17	00
30 KVA - INDOOR 3φ DRY XFMR	30	150°C	H-5	1	—		1100	00
200a - 4W/SN FUSIBLE DISC. SW.	31		B-16	1	—		420	00
1 1/2" EMT	32		E-1	15'	2	50	37	50
4 #1 THW, CU	33		E-8	15'	4	50	67	50
INSTALL 20 HP STARTER	34	FURNISHED BY OTHERS	G-17	1	—		150	00
PNL. "B" 208V., 30 CCT., 3φ MLO	35		B-11	1	—		900	00
PNL. "A" 277/480V. 42 CCT. MLO	36		B-13	1	—		1700	00
1 #6 BARE GROUND (COPPER)	37		D-2	25'	1	10	27	50
APPLIANCE CONNECTION - 60a	38	3 WIRE	F-24	1	—		27	00
APPLIANCE CONNECTION - 125a	39	3 WIRE + 10% FOR NEUT	F-24	1	—		54	50

DESCRIPTION	QTY.	UNIT LABOR		EXTENDED LABOR HOURS		UNIT MAT'L.		EXTENDED MAT'L. $		210	04
PHASE LOSS RELAY	1	2	00	2	00	—		137	00	6474	14
										SHEET TOTAL (Unadjusted)	

Total labor hours		2	00
Labor rate per hour		× 16.00	
Total labor dollars		32	00
Multiplier		× 1.97	
Total labor, overhead, & direct job expenses		63	04
Total material cost		137	00
Total gross cost		200	04	× 1.05 =

REEP

SUMMARY SHEET

JOB _Example #1 – Single Line Diagram_   DATE _4/2/85_

LOCATION_____

ARCHITECT_____ ORIGINAL PRICE____

ENCLOSED AREA_____ $/sq ft._____ CHANGE ORDER____

CONNECTED LOAD (W/sq.ft.)_____ DIVERSIFIED DEMAND LOAD (W/sq.ft.)_____

ALTERNATES_____

DESCRIPTION OF WORK _SERVICE EQUIPMENT, POWER DISTRIBUTION_
          AND PANELBOARDS

REMARKS_____

	SHEET NO	DESCRIPTION	AMOUNT	
	1	SINGLE LINE DIAGRAM	13,263	70
	2	SINGLE LINE DIAGRAM	6,474	14
		Total Summarized Price	19,737	84
		Price Adjustment Multiplier	X 1.34	
		Adjusted Sell Price	26,448	71

PRICING SHEET TOTALS

	DESCRIPTION	AMOUNT		
	Total of subcontract items			
	Multiplier	X 1.05		
	Adjusted total			

SUBCONTRACT ITEMS

TOTAL JOB PRICE ⟶ | 26,448 | 71

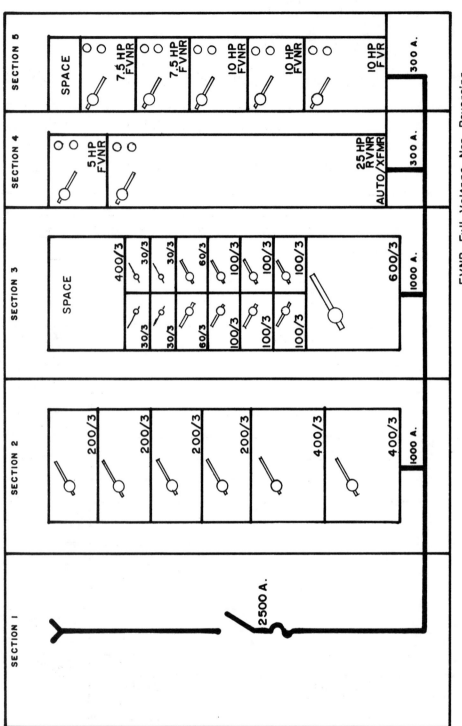

FVNR – Full Voltage Non – Reversing
RVNR – Reduced Voltage Non – Reversing
FVR – Full Voltage Reversing

120/208 VOLT MAIN DISTRIBUTION PANEL & MOTOR CONTROL CENTER

EXAMPLE NO. 2

REEP
PRICING SHEET

JOB _EXAMPLE #2_   SHEET NO. _1_

WORK _120/208 V. MAIN DISTR. PANEL & MTR. CONTROL CNTR._ OF _1_

PRICED BY _____   DATE _4/2/85_

DESCRIPTION	REF PAGE	REMARKS or DEVIATIONS	QUANTITY	UNIT PRICE	EXTENDED PRICE
2500a BOLTED PRESSURE SWITCH	B-1		1	—	9000 00
COST OF CBT. SECTIONS 2 & 3	B-2	3φ-4W MLO	2	2300 00	4600 00
600a BRANCH SWITCH UNIT	B-5		1	—	1600 00
400a BRANCH SWITCH UNIT	B-5		2	1125 00	2250 00
200a BRANCH SWITCH UNIT	B-5		4	560 00	2240 00
100a TWIN BRANCH UNIT	B-5		3	410 00	1230 00
60a TWIN BRANCH SWITCH	B-5		1	—	285 00
30a TWIN BRANCH SWITCH	B-5		2	260 00	520 00
COST OF MCC CBT SECTION 4 & 5	G-25	600a 3φ-3W	2	1150 00	2300 00
5 HP FULL VOLTAGE NON-REV. STR.	G-28	SIZE 1	1	—	840 00
7 1/2 HP FULL VOLTAGE NON-REV. STR.	G-28	SIZE	2	840 00	1680 00
10 HP FULL VOLTAGE NON-REV. STR.	G-28	SIZE 2	2	1030 00	2060 00
10 HP FULL VOLTAGE REVERSING STR.	G-29	SIZE 2	1	—	1150 00
25 HP AUTO XFMR. RED. VOLT. STR.	G-31	SIZE 3	1	—	2250 00

DESCRIPTION	QTY.	UNIT LABOR	EXTENDED LABOR HOURS	UNIT MAT'L.	EXTENDED MAT'L. $	
						→ 32,005 00
						SHEET TOTAL (Unadjusted)

Total labor hours			
Labor rate per hour		× 16.00	
Total labor dollars			
Multiplier		× 1.97	
Total labor, overhead, & direct job expenses			
Total material cost			
Total gross cost ———————→		———→ × 1.05 = —	

REEP

SUMMARY SHEET

JOB _EXAMPLE #2 - MAIN DISTR PNL & MTR. CONTR. CNTR._ DATE _4/2/85_

LOCATION _____

ARCHITECT _____ ORIGINAL PRICE _____

ENCLOSED AREA _____ $/sq ft. _____ CHANGE ORDER _____

CONNECTED LOAD (W/sq ft) _____ DIVERSIFIED DEMAND LOAD (W/sq.ft.) _____

ALTERNATES _____

DESCRIPTION OF WORK _3 SECTION FUSIBLE DISTRIBUTION PANEL AND_
_____ 2 SECTION MOTOR CONTROL CENTER - ALL 208V. - 3φ_

REMARKS _____

	SHEET NO	DESCRIPTION	AMOUNT	
PRICING SHEET TOTALS	1	DISTRIBUTION PANEL & M.C.C.	32,005	00
		Total Summarized Price	32,005	00
		Price Adjustment Multiplier	× 1.34	
		Adjusted Sell Price	42,887	—

	DESCRIPTION	AMOUNT		
SUBCONTRACT ITEMS				
	Total of subcontract items			
	Multiplier	×	1.05	
	Adjusted total			

TOTAL JOB PRICE ——→ 42,887 —

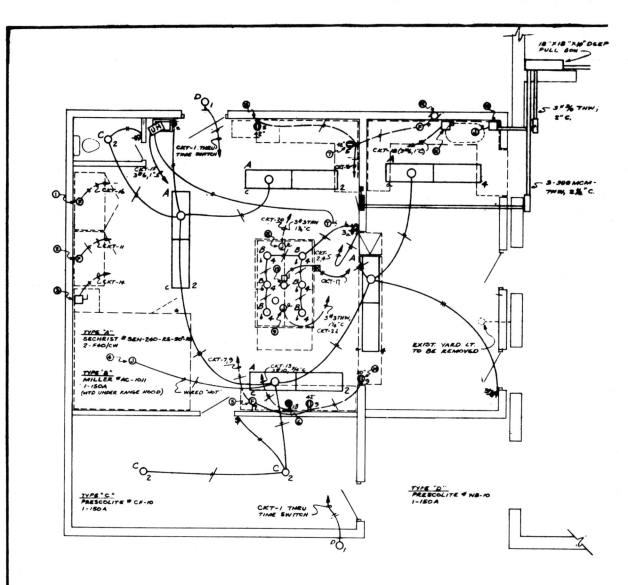

ELECTRICAL PLAN
SCALE: ⅛"-1'-0"

	KITCHEN EQUIPMENT SCHEDULE			
MARK	DESIGNATION	HP OR KW	VOLTS	CONTRACTOR TO FURNISH
①	REACH IN FREEZER	¾ HP	240	BUSS #SSY FUSEHOLDER & SWITCH W/ FLEX CONNECTION 12" UP.
②	REACH IN REFRIGERATOR	⅓ HP	120	BUSS #SSU FUSEHOLDER & FLEX CONN. 12" UP
③	WALK-IN REFRIGERATOR	1 HP	240	30a, 2P DISC. SWITCH & FLEX CONN. 36" UP.
④	LTG OUTLET FOR REFRIG.	.1 KW	120	J BOX 7' UP.
⑤	MIXER	½ HP	120	BUSS SSU FUSEHOLDER & FLEX CONN. 12" UP
⑥	FUTURE TOASTER	5 KW	240	30a, 250V REC. 48" UP.
⑦	GEN. USE RECEPTACLE	.2 KW	120	REC. MTD. 42" UP.
⑧	GRIDDLE & OVEN	22 KW	240	J. BOX & FLEX CONNECTION 6" UP
⑨	GRIDDLE & OVEN	22 KW	240	J. BOX & FLEX CONNECTION 6" UP
⑩	BOOSTER HEATER	36 KW	240	RELOCATED 200a, 2P DISC. SW, & FLEX, CONNECTION 12" UP
⑪	DISHWASHER	10 KW	240	60a, 2P No FUSE DISCONNECT SWITCH & FLEX CONNECTION 12" UP
⑫	EXHAUST FAN	⅟₂₀ HP	120	BUSS #SSU & FLEX CONNECTION 7' UP.
⑬	EXHAUST FAN	¾ HP	240	30a, 2P NO FUSE RAINTITE DISC. SWITCH & MANUAL MOT. STARTER
⑭	REC. FOR FUT. HOT PANS	.7 KW	120	REC. 30" UP
⑮	PEELER	⅛ HP	120	REC. 42" UP.

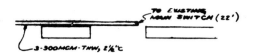

TO EXISTING
MAIN SWITCH (22')

3-300MCM-TNW, 2½"C

EXAMPLE NO. 3

SCHEDULE FOR PANEL "A"

120/240 VOLTS 1 PHASE 3 WIRE

FLUSH MTG. MAINS MLO

NOTES SQ. D #NQOB-42-3L
(OR EQUAL)

LOAD	CKT. NO.	BKR.	φ	BKR.	CKT. NO.	LOAD
EXT. LTG	1	20/1	A	20/1	2	LTG
SPARE	3		B		4	
PEELER & REC.	5		A		6	SPARE
MIXER	7		B		8	SPARE
REC.	9		A		10	SPARE
REFRIG	11		B		12	SPARE
TOASTER	13	30/2	A	20/2	14	WALK-IN REFRIG
			B			
UNIT HTR	15	50/2	A	20/2	16	FREEZER
			B			
EXH. FAN	17	20/2	A	70/2	18	DISHWSHR
			B			
SPARE	19	30/2	A	100/2	20	GRIDDLE
			B			
—	—	—	A	100/2	22	GRIDDLE
—	—	—	B			
SPACE	—	—	—	—	—	SPACE
	—	—	—	—	—	

SYMBOL LIST

SYMBOL	DESCRIPTION
	FLUORESCENT OR INCANDESCENT FIXTURE - UPPER CASE LETTER DENOTES TYPE, NO. DENOTES CIRCUIT NO. LOWER CASE LETTER, CONTROLLING SWITCH
	15A, 1P SWITCH. MTG. HT. - 48" ABOVE FLOOR. ARROW HART #1101-I & IVORY PLATE
	15A, 3 WAY SWITCH MTG. HT. - 48" ABOVE FLOOR. A.H. #1103-I & IVORY PLATE
	15A, DUPLEX CONVENIENCE OUTLET MTG. HT. AS SHOWN ON PLAN. A.H. #5651-I & IN. PL.
	30A, 250V REC - ARROW HART #9344-I & IVORY PLATE
	BUSS FUSEHOLDER OF TYPE INDICATED ON EQUIPMENT SCHEDULE
	JUNCTION BOX.
	UNIT HEATER - CHROMALOX #LUN-100, 240V, 1φ WITH #CR 360 CONTACTOR.
	THERMOSTAT - CHROMALOX #MNT 498A. MTD. 5'-0" UP.
	TIMESWITCH - TORK #T2120L FL-5
	MANUAL MOTOR STARTER MTD. ON HOOD
	BRANCH CIRCUIT CONDUIT RUN CONCEALED ABOVE CEILING OR IN FLOOR
	BRANCH CIRCUIT CONDUIT RUN CONCEALED IN FLOOR.

KITCHEN ADDITION
CAMP ALEXANDER
PARK COUNTY, COLORADO
ELEVEN-MILE RESERVOIR

E. KENNETH KOLSTAD
2714
STATE OF COLORADO

CONSULTING ENGINEER

JOB NO:
DATE:
SCALE
DRAWN:
CHECKED:
APPROVED
APPROVED

E-1

REEP
PRICING SHEET

JOB _EXAMPLE #3 - CAMP ALEXANDER - B.S.A._   SHEET NO. _1_

WORK _INTERIOR ELECTRICAL_   OF _2_

PRICED BY _G.V.K._   DATE _4/3/85_

DESCRIPTION	REMARKS or DEVIATIONS	REF. PAGE	QUANTITY	UNIT PRICE		EXTENDED PRICE	
TYPE A LIGHTING FIXTURE		A-7	10	123	00	1230	00
TYPE B " "		A-17	6	41	00	246	00
TYPE C " "	USED WB-23 COST	A-15	3	65	00	195	00
TYPE D " "	DITTO	A-15	2	65	00	130	00
120/240 1Ø-3W. BR. PNL 42 CCT		B-10	1	—		1100	00
30/2, NO FUSE DISC. SWITCH	USED FUSIBLE SWITCH COST	B-14	1	—		78	00
60/2, " " "	"	B-14	1	—		100	00
30/2, " RAINTITE DISC. SW	"	B-14	1	—		102	00
2½" GRC CONDUIT		E-1	70'	5	50	385	00
2" GRC CONDUIT		E-1	15'	4	15	62	25
1¼" " "		E-1	35'	3	10	108	50
1" " "		E-1	45'	2	25	101	25
2½" LB CONDULET		E-3	1	—		123	00
2" LB CONDULET		E-3	1	—		84	00
3#300 MCM THW CONDUCTOR		E-8	73'	9	80	715	40
3#3/0 THW		E-8	21'	6	00	126	00
3#3 THW "		E-8	40'	2	65	106	00
3#6 THW "		E-8	50'	1	75	87	50
300 MCM TERM. IN P.B. & EXIST. SW.		E-16	6	40	00	240	00
2#12, ½" EMT		F-2	395'	1	90	150	50
3#12, ½" EMT		F-2	80'	2	05	164	00
4#12, ½" EMT		F-2	10'	2	20	22	00
3#10, ½" EMT		F-2	30'	2	30	69	00

DESCRIPTION	QTY.	UNIT LABOR	EXTENDED LABOR HOURS	UNIT MAT'L.	EXTENDED MAT'L. $		
						6325	40
						SHEET TOTAL (Unadjusted)	

Total labor hours			
Labor rate per hour	× 16.00		
Total labor dollars			
Multiplier	× 1.97		
Total labor, overhead, & direct job expenses			
Total material cost			
Total gross cost		× 1.05 =	

REEP

PRICING SHEET

JOB _EXAMPLE #3- CAMP ALEXANDER - B.S.A._ SHEET NO. _2_

WORK _INTERIOR ELECTRICAL_ OF _2_

PRICED BY _G.V.K._ DATE _4/3/85_

DESCRIPTION	REMARKS or DEVIATIONS		QUANTITY	UNIT PRICE		EXTENDED PRICE	
CLG. FIXTURE OUTLETS & J-BOXES		F-18	10	27	00	270	00
WALL FIXTURE OUTLETS		F-18	2	22	50	45	00
15a DUPLEX RECEPT. OUTLETS		F-19	4	41	00	164	00
1P WALL SWITCH		F-22	4	38	50	154	00
3 WAY WALL SWITCH		F-22	2	43	50	87	00
2 WIRE THERMOSTAT		J-15	1	—		52	25
TIME SWITCH W/ASTRO. DIAL	W/CARRY OVER	G-5	1	—		283	25
30a, 250V. RECEPTACLE		F-19	1	—		47	25
BUSS FUSEHOLDER & FLEX. CONN.		G-16	4	73	75	295	00
3/4 HP MANUAL MOTOR STARTER		G-16	1	—		73	75
200a-2W APPLIANCE CONNECTION		F-24	1	—		64	00
100a-2W '' ''		F-24	4	34	00	136	00
60a-2W '' ''		F-24	1	—		23	00
18"x18"x10" PULL BOX	324 SQ. IN.	L-1	1	—		120	00
6"x 6"x 4" JUNCTION BOX	36 SQ. IN.	L-1	2	46	00	92	00
2½" HOLES IN P.B. & EXIST. SWITCH		L-6	3	9	70	29	10
2" HOLE IN PULLBOX		L-6	1	—		8	25
RELOCATE EXISTING SWITCH	LESS SW. COST	B-13	1	—		204	00
10 KW HEATER		J-1	1	—		490	00

DESCRIPTION	QTY.	UNIT LABOR	EXTENDED LABOR HOURS	UNIT MAT'L.	EXTENDED MAT'L. $	
						2637 | 85
						SHEET TOTAL (Unadjusted)

Total labor hours			
Labor rate per hour		× 16.00	
Total labor dollars			
Multiplier		× 1.97	
Total labor, overhead, & direct job expenses			
Total material cost			
Total gross cost			× 1.05 =

REEP

SUMMARY SHEET

JOB _Example #3 - Camp Alexander_   DATE _4/3/85_

LOCATION _____

ARCHITECT _____   ORIGINAL PRICE ____

ENCLOSED AREA _____ $/sq ft. _____   CHANGE ORDER ____

CONNECTED LOAD (W/sq ft.) _____   DIVERSIFIED DEMAND LOAD (W/sq.ft.) _____

ALTERNATES _____

DESCRIPTION OF WORK _INTERIOR ELECTRICAL - COMPLETE_ ___

REMARKS _____

	SHEET NO	DESCRIPTION	AMOUNT	
	1	INTERIOR ELECTRICAL	6,325	40
	2	INTERIOR ELECTRICAL	2,637	85
		Total Summarized Price	8,963	75
		Price Adjustment Multiplier	X 1.34	
		Adjusted Sell Price	12,010	76

PRICING SHEET TOTALS

	DESCRIPTION	AMOUNT			
	Total of subcontract items				
	Multiplier	X	1.05		
	Adjusted total				

SUBCONTRACT ITEMS

TOTAL JOB PRICE ⟶ 12,010 | 76

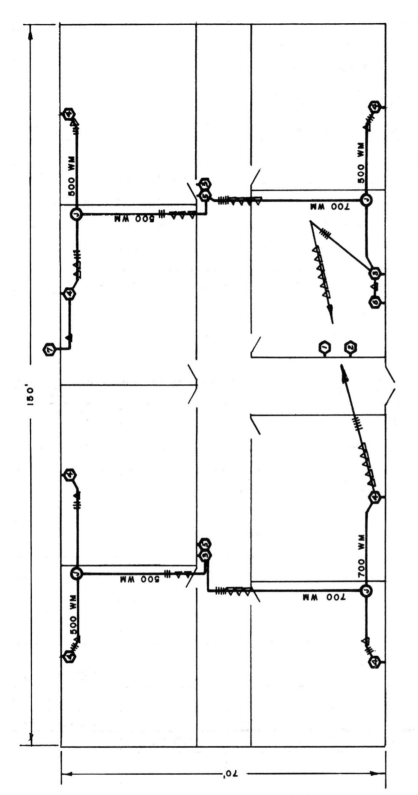

INSTALLATION OF CLOCK & SOUND SYSTEM IN
EXISTING OLDER SCHOOL

EXAMPLE NO. 4

① SYNCHRONOUS MASTER CLOCK with program control & daylight savings feature

② COMPACT COMMUNICATION CENTER

③ SPEAKER, surface type

④ TIME-TONE UNIT wy 12" sq. corrected clock & 8" speaker

⑤ PROGRAM BELL – 6"

⑥ VOLUME CONTROL

⑦ EXTERIOR HORN – 15w

━╫━△△△━ WIREMOLD in size indicated with 3-14 & sound cables indicated by triangles

REEP
PRICING SHEET

JOB _EXAMPLE #4 - EXISTING SCHOOL BUILDING_   SHEET NO. _1_

WORK _INSTALLATION OF CLOCK & SOUND SYSTEM_   OF _1_

PRICED BY   DATE _4/3/85_

DESCRIPTION	KEY	REMARKS or DEVIATIONS	REF PAGE	QUANTITY	UNIT PRICE		EXTENDED PRICE	
SYNCH. WIRED MASTER CLOCK	1	SURF. MTD.	I-11	1	—		2913	00
COMPACT COMMUNICATION CENTER	2		I-24	1	—		1813	00
DESK MIKE	—		I-27	1	—		94	00
SURFACE MTD SPEAKER	3		I-25	3	208	00	624	00
EXTERIOR HORN	7		I-25	1	—		143	00
TIME TONE UNIT FOR 12" CLOCK	4	SURF MTD	I-11	7	122	00	854	00
12" SYNC. WIRED CLOCK	4		I-10	7	162	00	1134	00
8" SPEAKER & XFMR	4		I-25	7	33	00	231	00
6" PROGRAM BELL	5		I-9	2	104	00	208	00
SURF. MTD. VOLUME CONTROL	6		I-25	1	—		59	00
WIREMOLD J-BOX	—		F-29	4	21	50	86	00
500 WIREMOLD	—		F-28	190'	1	90	361	00
700 "	—		F-28	180'	1	95	351	00
3#14 PULLED IN	—		F-9	240'		63	151	20
5#14 PULLED IN	—		F-9	200'		83	166	00
2/c #20 SPEAKER CABLE	—		I-28	1175'		59	693	25
DRILL 1" HOLE IN MASONRY WALL	—		L-3	10	20	00	200	00

DESCRIPTION	QTY.	UNIT LABOR	EXTENDED LABOR HOURS	UNIT MAT'L.	EXTENDED MAT'L. $		
						10,081	45
						SHEET TOTAL (Unadjusted)	

Total labor hours			
Labor rate per hour		× 16.00	
Total labor dollars			
Multiplier		× 1.97	
Total labor, overhead, & direct job expenses			
Total material cost			
Total gross cost			× 1.05 =

REEP

SUMMARY SHEET

JOB _EXAMPLE #4 - EXISTING SCHOOL, CLOCK & SOUND SYS._ DATE _4/3/85_

LOCATION _____

ARCHITECT _____ ORIGINAL PRICE ____

ENCLOSED AREA _____ $/sq ft. _____ CHANGE ORDER ____

CONNECTED LOAD (W/sq.ft.) _____ DIVERSIFIED DEMAND LOAD (W/sq.ft.) _____

ALTERNATES _____

DESCRIPTION OF WORK _INSTALLATION OF CLOCK & SOUND SYSTEM IN_
EXISTING OLDER SCHOOL

REMARKS _____

	SHEET NO	DESCRIPTION	AMOUNT	
PRICING SHEET TOTALS	1	CLOCK & SOUND SYSTEM INCLUDING WIRING	10,081	45
		Total Summarized Price	10,081	45
		Price Adjustment Multiplier	X 1.34	
		Adjusted Sell Price	13,509	14

	DESCRIPTION	AMOUNT		
SUBCONTRACT ITEMS				
	Total of subcontract items			
	Multiplier	X 1.05		
	Adjusted total			

TOTAL JOB PRICE ⟶ | 13,509 | 14 |

REEP

SUMMARY SHEET

JOB _____ DATE _____

LOCATION _____

ARCHITECT _____ ORIGINAL PRICE ____

ENCLOSED AREA _____ $/sq ft. _____ CHANGE ORDER ____

CONNECTED LOAD (W/sq ft.) _____ DIVERSIFIED DEMAND LOAD (W/sq.ft.) _____

ALTERNATES _____

_____

DESCRIPTION OF WORK _____

_____

_____

REMARKS _____

	SHEET NO.	DESCRIPTION	AMOUNT	
PRICING SHEET TOTALS				
	Total Summarized Price			
	Price Adjustment Multiplier	×		
	Adjusted Sell Price			

	DESCRIPTION	AMOUNT		
SUBCONTRACT ITEMS				
	Total of subcontract items			
	Multiplier	×	1.05	
	Adjusted total			

TOTAL JOB PRICE ⟶

REEP

SUMMARY SHEET

JOB_____ DATE_____

LOCATION_____

ARCHITECT_____ ORIGINAL PRICE____

ENCLOSED AREA_____ $/sq.ft._____ CHANGE ORDER____

CONNECTED LOAD (W/sq ft)_____ DIVERSIFIED DEMAND LOAD (W/sq.ft.)_____

ALTERNATES_____

_____

DESCRIPTION OF WORK_____

_____

_____

REMARKS_____

SHEET NO	DESCRIPTION	AMOUNT	
	Total Summarized Price		
	Price Adjustment Multiplier	×	
	Adjusted Sell Price		

PRICING SHEET TOTALS

DESCRIPTION	AMOUNT			
Total of subcontract items				
Multiplier	×	1.05		
Adjusted total		→		

SUBCONTRACT ITEMS

TOTAL JOB PRICE ——→

REEP
PRICING SHEET

JOB _____   SHEET NO. _____
WORK _____   OF _____
PRICED BY _____   DATE

DESCRIPTION	REMARKS or DEVIATIONS	QUANTITY	UNIT PRICE		EXTENDED PRICE	

DESCRIPTION	QTY.	UNIT LABOR	EXTENDED LABOR HOURS	UNIT MAT'L.	EXTENDED MAT'L. $		

SHEET TOTAL (Unadjusted)

Total labor hours		
Labor rate per hour	× 16.00	
Total labor dollars		
Multiplier	× 1.97	
Total labor, overhead, & direct job expenses		
Total material cost		
Total gross cost ———————————→		→ × 1.05 = —

REEP

PRICING SHEET

JOB _____   SHEET NO. _____

WORK _____   OF _____

PRICED BY _____   DATE

DESCRIPTION	REMARKS or DEVIATIONS	QUANTITY	UNIT PRICE		EXTENDED PRICE	

DESCRIPTION	QTY.	UNIT LABOR	EXTENDED LABOR HOURS		UNIT MAT'L.	EXTENDED MAT'L. $			
									SHEET TOTAL
									(Unadjusted)

Total labor hours		
Labor rate per hour	× 16.00	
Total labor dollars		
Multiplier	× 1.97	
Total labor, overhead, & direct job expenses		
Total material cost		
Total gross cost ———————→		——→ × 1.05 = —

REEP
PRICING SHEET

JOB _____   SHEET NO. _____

WORK _____   OF _____

PRICED BY _____   DATE

DESCRIPTION	REMARKS or DEVIATIONS	QUANTITY	UNIT PRICE		EXTENDED PRICE	

DESCRIPTION	QTY.	UNIT LABOR	EXTENDED LABOR HOURS		UNIT MAT'L.	EXTENDED MAT'L. $			
								SHEET TOTAL (Unadjusted)	

Total labor hours			
Labor rate per hour	× 16.00		
Total labor dollars			
Multiplier	× 1.97		
Total labor, overhead, & direct job expenses			
Total material cost			
Total gross cost		× 1.05 =	

REEP

PRICING SHEET

JOB _____ SHEET NO. _____

WORK _____ OF _____

PRICED BY _____ DATE

DESCRIPTION	REMARKS or DEVIATIONS	QUANTITY	UNIT PRICE	EXTENDED PRICE

DESCRIPTION	QTY.	UNIT LABOR	EXTENDED LABOR HOURS	UNIT MAT'L	EXTENDED MAT'L $	
						SHEET TOTAL (Unadjusted)

Total labor hours	
Labor rate per hour	X 16.00
Total labor dollars	
Multiplier	X 1.97
Total labor, overhead, & direct job expenses	
Total material cost	
Total gross cost	X 1.05 =

REEP

PRICING SHEET

JOB_____ SHEET NO._____

WORK_____ OF_____

PRICED BY_____ DATE

DESCRIPTION	REMARKS or DEVIATIONS	QUANTITY	UNIT PRICE	EXTENDED PRICE	

DESCRIPTION	QTY.	UNIT LABOR	EXTENDED LABOR HOURS	UNIT MAT'L	EXTENDED MAT'L $	
						SHEET TOTAL (Unadjusted)

Total labor hours		
Labor rate per hour	X 16.00	
Total labor dollars		
Multiplier	X 1.97	
Total labor, overhead, & direct job expenses		
Total material cost		
Total gross cost ─────────────►		X 1.05 =

REEP
PRICING SHEET

JOB _____

WORK _____

PRICED BY _____

SHEET NO. _____

OF _____

DATE

DESCRIPTION	REMARKS or DEVIATIONS	QUANTITY	UNIT PRICE		EXTENDED PRICE	

DESCRIPTION	QTY.	UNIT LABOR		EXTENDED LABOR HOURS		UNIT MAT'L		EXTENDED MAT'L $			
											SHEET TOTAL (Unadjusted)

Total labor hours			
Labor rate per hour	× 16.00		
Total labor dollars			
Multiplier	× 1.97		
Total labor, overhead, & direct job expenses			
Total material cost			
Total gross cost ——————————▶		————▶ × 1.05 =	

Index

Anchors (*see* Fastenings)
Antenna, TV (*see* Master antenna, TV)
Appliance connections, F-24
Arresters, lightning, D-10, K-8
Asphalt-pavement cutting, L-7

Bare copper grounding conductor, D-2
Bells (*see* Signaling)
Bollard lights, A-36
Bolted pressure switches, B-1
Boxes, junction (*see* Junction boxes)
Busway:
　feeder, copper and aluminum, E-18
　　elbows, copper and aluminum, E-20
　　tees, copper and aluminum, E-21
　　terminals, copper and aluminum, E-22
　　transformer taps, copper and aluminum, E-19
　plug-in, copper and aluminum, E-23
　　cable tap box, copper and aluminum, E-26
　　elbows, copper and aluminum, E-24
　　switches, 250 and 600 volt, E-27
　　tees, copper and aluminum, E-25
Buzzers (*see* Signaling)

Cabinets:
　current transformer, C-2
　distribution panelboard, B-2 to B-4, B-6, B-7
　telephone, I-1
Cable:
　armored, F-14
　lightning protection, D-5
　low-voltage: high-temperature signaling and power-limited, I-29
　　remote control, G-3
　　TV, I-35
　mat-type asphalt/concrete heating, J-12–J-13

Cable (*Contd.*):
　mineral-insulated snow-melting, jacketed, J-14
　mineral-insulated wiring, F-15, F-16
　nonmetallic sheathed, F-11, F-13
　roof and pipe heating, J-11
　service drop, self-supporting, E-13
　service entrance, E-12
　telephone, high-temperature, I-5
　undercarpet (*see* Undercarpet wiring system)
　underground, 15 kV, URD, K-7
　underground feeder, type UF, F-13
Cadweld connections (welded), D-3, D-4
Capacitors, power-factor corrective, G-23
Channeling, concrete and masonry, L-6
Chimes (*see* Signaling)
Circuit breakers:
　for distribution panelboards, B-8
　interrupting ratings, B-20
　for lighting/appliance panelboards, B-12
　separately enclosed: 15 to 60 ampere, B-20
　　70 to 100 ampere, B-21
　　125 to 225 ampere, B-22
　　250 to 400 ampere, B-23
　　450 to 600 ampere, B-24
　　700 to 800 ampere, B-25
　　900 to 1000 ampere, B-26
Clamps, ground, D-2, D-9
Clock/program system:
　accessories: buzzers, I-12
　　clock guards, I-12
　　coded relays, I-12
　　clocks (indicating, synchronous wired, and electronic), I-10
　　frequency generators, I-12
　　master time/program center, I-11
　　time-tone units, I-11
Compression terminals, E-17
Concrete-pavement cutting, L-8
Conductors: bare ACSR, K-6
　branch circuit, all copper: armored cable, F-14

Conductors, branch circuit, all copper (*Contd.*):
　bare copper, D-2
　mineral-insulated wiring cable, F-15, F-16
　nonmetallic sheathed cable, F-11
　TFF/THHN, F-9
　thermostat/annunciator, F-10
　XLPE/USE, direct-burial, F-12
　feeder: aluminum: insulated grounding, E-16
　　service drop, E-13
　　service entrance, E-12
　　USE/XLPE, direct-burial, E-15
　　XHHW and THW, E-14
　copper: bare, D-2
　　insulated grounding, E-16
　　service entrance, E-12
　　THW insulation, E-8
　　THWN/THHN insulation, E-10
　　USE/XLPE, direct-burial, E-11
　　XHHW insulation, E-9
Conduit fittings:
　couplings (EMT, ARC, GRC, PVC, and Erickson), E-4
　elbows, factory-made, E-7
　field bends, E-6
　nipples (straight, offset, and chase), E-5
　seal-offs (LB, LL, LR, LF, CO, and T), E-3
　terminals, E-2
Conduits:
　empty: aluminum rigid conduit (ARC), E-1
　　electrical metallic tubing (EMT), E-1, F-5
　　flexible metal (FLEX), E-1, F-5
　　galvanized rigid conduit (GRC), E-1, F-5
　　intermediate metal conduit (IMC), E-1, F-5
　　nonmetallic FLEX, F-6
　　nonmetallic tubing, F-6
　　polyvinyl chloride (PVC), E-1, F-5

Conduits, empty (*Contd.*):
  PVC-coated galvanized rigid metal, E-1, F-5
  with wire: EMT with copper TFF, F-1
    EMT with copper THWN/THHN, F-2
    GRC/IMC with copper TFF, F-3
    PVC with copper TFF or THHN, F-4
Connections:
  appliance, F-24
  motor terminal, F-25, F-26
Connectors:
  compression type, E-17
  split-bolt type, E-17
Contactors:
  electrically held, 30 to 300 ampere, G-8
  mechanically held, 30 to 300 ampere, G-9
Current transformer cabinet, C-2
Cutouts, fused, high-voltage, K-8
Cutting pavement:
  asphalt, L-7
  concrete, L-8

Detectors (*see* Fire alarm system)
Dimmers:
  fluorescent: manual remote, G-14
    motorized remote, G-15
    wall type, manual, G-13
  incandescent: manual remote, G-11
    motorized remote, G-12
    wall type, manual, G-10
Distribution panelboard:
  cable interconnect, B-9
  circuit-breaker type: branch circuit breakers, B-8
    cabinet only: with main circuit breaker, B-7
      with main lugs only, B-6
  fusible type: branch circuit switches, B-5
    cabinet only: with main lugs only, B-2
      with 240-volt main switch, B-3
      with 600-volt main switch, B-4
Door lock release, I-30
Door switches, F-22
Drilling holes:
  in concrete, L-4
  in masonry, L-3
  in steel panel, L-5
Drop cable, self-supporting, E-13
Dryer outlet, F-19
Duct banks:
  concrete-encased, plastic duct, E-29
  direct burial, earth cover, plastic duct, E-28
  reinforced-concrete-encased, plastic duct, E-30
Ducts:
  lighting (*see* Lighting duct)
  trench (*see* Trench duct)

Ducts (*Contd.*):
  underfloor (*see* Underfloor duct)
Duplex receptacles, F-19

Electric heaters (*see* Heaters)
Electrical metallic tubing, E-1, F-5
Emergency AC power systems, battery-powered, A-19, A-20
Emergency lighting, battery-powered, A-18, A-20
Emergency telephones, I-18, I-31
Entrance caps (weatherhead), C-1, E-3
Erickson coupling, E-4
Exit lighting:
  non-self-powered, A-16
  self-powered, A-20
Exothermic (welded) connections:
  cable-to-cable, D-3
  cable-to-steel, D-4

Fastenings:
  to concrete, L-9, L-10
  to drywall, L-9
  to hollow masonry, L-9, L-10
  to metal, L-10
Feeder busway (*see* Busway, feeder, copper and aluminum)
Feeder conductors (*see* Conductors, feeder)
Fire alarm system:
  annunciator panel, I-15
  battery power pack, I-15
  cable, high-temperature, I-19
  control panel: standard, I-13
    with voice communication, I-14
  door holder and release, I-18
  fire detectors (*see* heat detector, *below*; smoke detector, *below*)
  heat detector, fixed temperature and rate of rise, I-16
  horn/flashing-light combination, I-18
  horn/speaker combination, I-18
  manual stations, I-16
  remote station receiving panel, I-15
  smoke detector: ionization type, I-17
    photoelectric type: ceiling-mounted, I-17, I-31
      duct-mounted, I-17
  sprinkler system alarm switches: OS&Y (valve position monitor) switch, I-16
    water-flow switch, I-16
  telephone, emergency, I-18
Fixture outlets, ceiling and wall, F-18
Fixtures (*see* Lighting fixtures)
Flat conductor cable system (*see* Undercarpet wiring system)
Floodlighting (*see* Lighting fixtures)
Floodlighting poles (*see* Lighting poles)
Fluorescent fixtures (*see* Lighting fixtures: exterior fluorescent; interior fluorescent)

Flushcalls (*see* Signaling)
Freeze protection, heat tape for, J-11
Fused cutouts, high-voltage, K-8

Generators, frequency, I-12
Ground clamps, D-2, D-9
Ground rods:
  for electrical service, D-1
  for lightning protection, D-9
Grounding conductors:
  bare, D-2
  insulated, E-16
Grounding connections, exothermic (welded):
  cable-to-cable, D-3
  cable-to-steel, D-4
Grounding plates, D-10
Gutter, service (*see* Wireway)

Heat tape for freeze protection, J-11
Heaters:
  commercial type: baseboard, J-4
    controls, J-15, J-16
    convectors: cabinet, J-8
      sill-height, J-5 to J-7
    infrared radiant, J-2
    mats, heating cable, J-12
    pipe heating cable, J-11
    snow-melting mats, J-13, J-14
    unit heaters, forced-air, J-1
    wall heaters, recessed with fan, J-9
  residential type: baseboard, J-3
    bathroom ceiling, J-10
    bathroom wall, J-10
    controls, J-15, J-16
    floor drop-in, J-10
    gutter and downspout, J-11
    kick space, fan forced, J-10
    radiant ceiling panels, J-10
    wall heaters, recessed with fan, J-9
Heating cable:
  floor-heating, J-12
  gutter and downspout, J-11
  pipe-heating, J-11
  snow-melting, J-13–J-14
High-intensity-discharge (HID) fixtures (*see* Lighting fixtures: exterior HID; interior HID)
High voltage (*see* Power distribution above 600 volts)
Hole drilling (*see* Drilling holes)
Horns (*see* Signaling)

Incandescent fixtures (*see* Lighting fixtures: exterior incandescent; interior incandescent)
Intercom system:
  apartment type: door lock release, I-30
    emergency call indicator, I-31
    emergency call switch, I-30

Intercom system, apartment type (*Contd.*):
  emergency telephone, I-31
  master control unit, I-31, I-33
  speaker unit, wall type, I-32
  small building type: call-in switch, I-24
    master unit, I-24
    speakers, I-25

Jacketed mineral-insulated snow-melting cables, J-14
Junction boxes, L-1
  underfloor duct, F-34

Ladder tray (*see* Trays, 4-inch-deep, steel)
Light control switches (*see* Switches)
Lighting contactors:
  electrically held, 30 to 300 ampere, G-8
  mechanically held, 30 to 300 ampere, G-9
Lighting duct:
  duct sections, F-38
  feed-in boxes, F-38
  receptacles and terminals, F-39
Lighting fixtures:
  exterior fluorescent, A-34, A-36
  exterior HID: bollards, A-36
    pole mount, A-33 to A-35
    surface mount, A-24, A-33, A-35
  exterior incandescent: bollards, A-36
    surface mount, A-24, A-33
  interior fluorescent: industrial, A-10
    strips, A-9
    surface mount, A-6 to A-10
    troffers: air-handling type, A-3 to A-5
      static type, A-1, A-2, A-8
    vandal-resistant, A-7
    vaportight, A-17
    wall mount, A-7
  interior HID: decorative shades, A-11
    enclosed and gasketed, A-11
    recessed and regressed, A-12, A-13
    surface mount, A-11
  interior incandescent: emergency type, A-18
    exit type, A-16, A-20
    explosion-proof, A-17
    industrial type, A-17
    Lite-Trac, A-16
    recessed, A-14
    residential type, A-21 to A-23
    surface mount, A-15, A-17
    vaportight, A-17
Lighting poles:
  aluminum: round tapered, 20 to 50 feet, A-26
    light-duty, 10 to 16 feet, A-29
    street lighting with pipe arms, A-30

Lighting poles, aluminum (*Contd.*):
    street lighting with truss arms, A-31
  steel: poletop brackets, A-32
    round tapered, 20 to 60 feet, A-25
    light-duty, 10 to 24 feet, A-29
    street lighting with pipe arms, A-30
    street lighting with truss arms, A-31
    square hinged, 20 to 40 feet, A-28
    square tapered, 20 to 45 feet, A-27
Lightning protection:
  air terminals, D-6
  arresters, D-10, K-8
  cables, D-5
  connectors, D-7 to D-9
Low-voltage remote control system:
  cables, G-3
  component cabinets, G-2
  relay, G-2
  switches, G-1, G-2
  system components, G-1 to G-3
  transformer, G-2

Master antenna, TV:
  amplifier, I-34
  cable, I-35
  roof antennas, I-34
  signal splitter, I-34
  wall outlet, I-35
Mat-type asphalt/concrete snow-melting cable, J-13
Metering, multi- (*see* Multimetering)
Mineral-insulated wiring cable, F-15, F-16
Motor control center:
  control devices, G-16
  feeder disconnects: circuit breakers, G-27
    fusible switches, G-26
  motor starters: control switches, G-16
    line voltage: full-voltage, nonreversing, G-28
    full-voltage, reversing, G-29
    full-voltage, two-speed, G-30
    reduced-voltage: autotransformer, G-31
    part-winding type, G-32
  vertical section: for incoming feeder, 20 inches deep, G-24
    for starters or disconnects, 20 inches deep, G-25
Motor control switches, pushbutton and selector type, G-16
Motor starters:
  combination full-voltage, G-18
  hand type for small motors, G-16
  line voltage, G-17
  reduced-voltage
    autotransformer, G-19
    part-winding type, G-21
    primary-resistor type, G-22
    wye-delta type, G-19

Motor terminal connections:
  115, 200, and 230 volt, F-25
  460 volt, F-26
Multimetering:
  main disconnect: circuit breaker, C-4
    switch, C-3
  single-phase group meters: with 10,000 A.I.R. circuit breakers, C-5
    with 22,000 A.I.R. circuit breakers, C-6
  three-phase group meters with 22,000 A.I.R. circuit breakers, C-7

Nipples, conduit, E-5
Nonmetallic sheathed cable, F-11, F-13
Nurse call system:
  bedside stations, I-21, I-22, I-30
  bell, corridor, I-30
  cables, I-23
  control station, I-20, I-22
  corridor dome light, I-23, I-30
  emergency call station, I-21
  power supply, I-21, I-22
  staff and duty stations, I-20, I-22

Outlets:
  ceiling, F-18
  floor, F-21
    poke-through type, F-23
    undercarpet wiring system, F-32
    underfloor duct, F-34
  telephone, F-21, F-23, F-34, I-2
  wall, F-18
    receptacles (*see* Receptacle outlets)
    switch, F-22
      dimmers, G-10, G-13
      door, F-22
    TV, I-35

Panelboard, distribution (*see* Distribution panelboard)
Panelboard circuit breakers, B-12
Panelboards with branch circuit breakers:
  with main lugs only: single-phase, 3 wire, 120/240 volt, commercial and residential, B-10
    three-phase, 4 wire, 120/208 volt, commercial and residential, B-11
    three-phase, 4 wire, 277/480 volt, commercial, B-12
Pavement cutting:
  asphalt, L-7
  concrete, L-8
Plug-in busway (*see* Busway, plug-in, copper and aluminum)
Poles:
  lighting (*see* Lighting poles)
  power distribution (*see* Power distribution above 600 volts)

Power distribution above 600 volts:
    overhead: bare ACSR conductor, K-6
        cutouts, arresters, and terminations,
        K-8
        poles, structures, supports, and
        accessories, K-1 to K-5
    underground: cable, 15 kV, aluminum
        URD, K-7
        terminations, K-8
Power-factor corrective capacitors, G-23
Price adjustment chart, xiv
Pushbuttons (*see* Signaling)

Raceways:
    surface: channel support and raceway
        system, F-27
        Wiremold, F-28
            fittings and devices, F-29
            telepower poles, F-30
        trench duct system, F-35 to F-37
        underfloor duct, F-33, F-34
Receptacle outlets:
    clock, F-21
    dryer, F-19
    duplex, single, and weatherproof
        duplex, F-19
    duplex with GFI, F-20
    floor, F-21, F-23
    lighting duct, F-39
    range, F-19
    undercarpet wiring, F-32
    underfloor duct, F-34
    weatherproof duplex, F-19
Relays:
    coded, I-12
    low-voltage remote control, G-2
    multipole, mechanically and electrically
        held, G-6
    quiet mercury type, 10 to 150 ampere,
        G-7
Remote control system (*see* Low-voltage
    remote control system)

Service drop cable, self-supporting, E-13
Service entrance cable, E-12
Service gutter (*see* Wireway)
Signaling:
    bells, I-7 to I-9, I-30
    buzzers, I-7, I-8, I-12
    cable, F-10, I-19, I-23
    chimes, I-7, I-9
    flushcalls, I-7
    horns, I-9, I-18
    pushbuttons, I-6, I-8, I-24
    sirens, I-9
    transformers, I-6, I-7, I-30, I-32

Sirens (*see* Signaling)
Snow detector, J-16
Snow-melting heating cables, J-13, J-14
Sound systems:
    amplifiers, I-26
    cables, I-23
    call-in and privacy switch, I-24
    master unit, I-24
    microphone receptacle, I-26
    microphones, I-27
    mixers, I-26
    speakers:
        auditorium, wide range, I-25
        trumpet reproducers, I-24
        wall and ceiling type, I-25
        volume controls, I-25
Starters, motor (*see* Motor starters)
Support system for fixtures, F-27
Surface raceways (*see* Raceways, surface)
Switches:
    bolted pressure, B-1
    control, pushbutton and selector type,
        G-16
    fusible safety: single-phase, 3 wire, 240
        volt, B-14
        three-phase, 3 wire, 240 volt, B-15
        three-phase, 4 wire, 240 volt, B-16
        three-phase, 4 wire, 480 volt, B-17
        three-phase, 3 wire, 600 volt, B-18
    low-voltage remote control, G-1, G-2
    molded case, 100 ampere, separately
        enclosed, B-27
    molded case, 225 to 1000 ampere,
        separately enclosed, B-28
    nonfusible safety, three-phase, 3 wire,
        250 and 600 volt, B-19
    photoelectric, G-4
    photo/time, G-4
    time, G-4, G-5
    timer, G-4

Telephone systems:
    cabinets, I-1
    cable, high-temperature, I-5
    emergency, I-18, I-31
    outlets, F-21, F-23, F-34, I-2
    undercarpet wiring: cable, I-3, I-4
        outlets, I-2
        transition boxes, I-2
Terminations:
    conductor, E-17
    high-voltage cable, K-8
    lighting duct, F-38
    mineral-insulated cable, F-17
Time switches:
    astronomic dial, G-5
    photo/time, G-4
    7-day dial, G-4

Time switches (*Contd.*):
    24-hour dial, G-5
    wall-box timer type, G-4
Time system (*see* Clock/program system)
Timer switch, G-4
Transformers:
    dry type:
        single phase: buck-boost, H-3
            general-purpose, H-1, H-2
            isolating-type, H-4
            low-voltage remote control, G-2
        three phase: autotransformer, H-6
            general-purpose, H-5
            isolating-type, H-7
    oil-filled: pad-mounted, three-phase, H-
        9
        pole-type, single-phase, H-8
Trays, ladder, 4-inch-deep, steel, E-31
    fittings: elbow: horizontal, E-31
            vertical, E-32
        reducer, straight, E-32
        tee, horizontal, E-32
Trench duct, F-35
    fittings: cross, F-37
        end closure, F-36
        horizontal elbow, F-37
        riser and cabinet connector, F-36
        tees, F-37
        vertical elbow, F-36
Trenching and backfilling, L-2
TV master antenna (*see* Master antenna,
    TV)

Undercarpet wiring system:
    flat conductor cable, F-31, I-3, I-4
        communications, I-3, I-4
        power cable, F-31
        taps and tees, F-32
    outlets: communications, I-2
        power, F-32
    transition box (round to flat wiring), F-
        32
Underfloor duct:
    fittings, F-34
    junction boxes, F-34
    outlets, power and telephone, F-34
    raceway, F-33
Underground cable, kV, URD, K-7
Underground feeder cable, type UF, F-13

Wall fastenings, L-9
Weatherhead (entrance caps), C-1, E-3
Wireway (service gutter):
    fittings (tees and elbows), C-9
    oiltight, C-10
    screw cover, C-8

ABOUT THE AUTHORS

C. Kenneth Kolstad holds a B.S. degree in electrical engineering from Northeastern University. Now retired, Mr. Kolstad was a member of the National Society of Professional Engineers, the Illuminating Engineering Society, the International Association of Electrical Inspectors, and the Consulting Engineers of Colorado, of which he is now a life member. He was formerly with the firm of Kolstad and Kohnert, Consulting Engineers in Colorado Springs.

Gerald V. Kohnert, P.E., is a graduate of the Illinois Institute of Technology. He is a certified consulting engineer and is also a member of the National Society of Professional Engineers, the Illuminating Engineering Society, the International Association of Electrical Inspectors, the Consulting Engineers of Colorado, and the Construction Specifications Institute. Mr. Kolstad and Mr. Kohnert have been associated professionally since 1960. Mr. Kohnert is carrying on their former partnership as Kohnert Engineering, Inc.